Biopolymer Electrolytes

Biopolymer Electrolytes

Fundamentals and Applications in Energy Storage

Y.N. Sudhakar

M. Selvakumar

D. Krishna Bhat

Elsevier
Radarweg 29, PO Box 211, 1000 AE Amsterdam, Netherlands
The Boulevard, Langford Lane, Kidlington, Oxford OX5 1GB, United Kingdom
50 Hampshire Street, 5th Floor, Cambridge, MA 02139, United States

Library of Congress Cataloging-in-Publication Data
A catalog record for this book is available from the Library of Congress

British Library Cataloguing-in-Publication Data
A catalogue record for this book is available from the British Library

ISBN: 978-0-12-813447-4

For information on all Elsevier publications visit our website at https://www.elsevier.com/books-and-journals

Publisher: Susan Dennis
Acquisition Editor: Anneka Hess
Editorial Project Manager: Michael Lutz
Production Project Manager: Nilesh Kumar Shah
Cover Designer: Vicky Pearson Esser

Typeset by SPi Global, India

Contents

Chapter 1

An introduction of Biopolymer Electrolytes

Chapter Outline

Biopolymer Electrolytes. https://doi.org/10.1016/B978-0-12-813447-4.00001-7

1.1 BIODEGRADABLE POLYMERS/BIOPOLYMERS

Biodegradable polymers/biopolymers are emerging as one of the hottest fields for addressing current environmental issues toward a sustainable future. This desire has made scientists explore natural polymers and mimic them with various combinations to synthesize them with better properties. They have also identified a few microorganisms and enzymes capable of degrading biopolymers. Explosive population growth has raised concerns in several parts of the world regarding issues such as deficiencies in food, resources, and energy as well as global environmental pollution. Science has to lead the world in a more mutual beneficial development by utilizing the lands in underdeveloped countries for growing the resources needed for biodegradable polymers. Dependence on synthetic polymers must decline because some countries are restricting the use of nonbiodegradable polymers. Synthetic polymers as of now are difficult to completely remove from the marketplace and may be produced until the fossil resources are available. Recycling of plastics is promoted more intensively nowadays, but recycling alone will not solve plastic pollution. Recycling requires considerable amounts of energy and eventually nonrecyclable plastics are incinerated or buried in landfills. Taking this into consideration, the importance and necessity of biodegradable polymers can be easily estimated [1,2]. The biodegradability of a polymer mainly depends on the chemical structure and products formed after biodegradation. Therefore, biopolymers are based on natural or synthetic materials.

Natural biopolymers are based mainly on renewable resources. Synthetic biopolymers usually are petroleum-based. To meet the functional requirements in the marketplace, many natural biopolymers are blended with synthetic polymers to get blended biopolymers. Having synthetic parts in the polymer chain makes the claim of biodegradability partially agreeable, as these are, in fact, bioerodable, photobiodegradable, or hydrobiodegradable. Along with microorganisms, environmental factors have an influence on the degradability of biopolymers. Nevertheless, biopolymers may be categorized based on their degradability in the environment under such terms as biodegradable, compostable, photobiodegradable, hydrobiodegradable, and bioerodable.

(A) BIODEGRADABLE

There are various definitions for biodegradation. One of them, according to ASTM, the biodegradable is defined as, "Capable of undergoing decomposition into carbon dioxide, methane, water, inorganic compounds, or biomass in which the predominant mechanism is the enzymatic action of microorganisms that can be measured by standardized tests, in a specified period of time reflecting available disposal conditions."

Biopolymers should be enzymatically broken down by microorganisms in a defined time into molecules such as carbon dioxide and water. The thickness of the biopolymer based on its fabrication and exposure to the environment highly influences the rate of biodegradation.

(B) COMPOSTABLE

Compostable is defined by ASTM as, "Capable of undergoing biological decomposition in a compost site as part of an available program, such that the plastic is not visually distinguishable and breaks down to carbon dioxide, water, inorganic compounds, and biomass, at a rate consistent with known compostable materials."

Compostable biopolymers must be able to biodegrade in the compost system within a specific time at slightly higher than the atmospheric temperature (typically around 12 weeks at temperatures over 50°C). The compost obtained after the biodegradation of biopolymer must not contain any distinguishable residue, heavy metal content, or ecotoxicity. Compostable biopolymers are a subset of biodegradable biopolymers (for example, cellulose).

(C) HYDROBIODEGRADABLE AND (D) PHOTOBIODEGRADABLE

Initially, the hydrolysis or photodegradation stage is involved in the breakdown of biopolymers. It is then followed by biodegradation of hydrobiodegradable and photobiodegradable polymers, respectively. Biopolymer which undergo degradation simultaneously by both hydrolysis and light also exist.

(E) BIOERODABLE

Environmental factors other than microorganisms that are able to degrade the biopolymers fall under the category of bioerodable biopolymers. This involves degradation of polymers as a dissolution in water, oxidative embrittlement, or photolytic embrittlement (ultraviolet (UV) aging).

1.1.1 Common Biopolymers

Starch is widely available, has a low cost, and is biodegradable; this means that starch is prevalent in materials of biodegradation interest such as carry bags, decorative articles, etc. Pure starch contains linear chain amylase, α-1,4 anhydroglucose units, and a highly branched amylopectin consisting of short chains linked by α-1,6 bonds. Nevertheless, pure starch is brittle and moisture-sensitive, thus strongly limiting its potential fields of application. It also has relatively poor mechanical properties and requires large amounts of plasticizers, such as glycerol or ethylene glycol, or requires modification of chemical properties of starch for preparation of films. Even blending with other polymers still showed a significant increase in the strength and flexibility of the starch biopolymer [3].

Polyester of natural origin that is produced by a wide variety of bacteria as intracellular reserve materials is receiving increased attention for possible applications as biodegradable biopolymers. Polyester can be molded to the desired shape because it can be melt-processed. Although aliphatic polyesters as high molecular weight, it undergoes biodegradation due to easily hydrolysable backbone which can fit into enzyme's active site while aromatic polyester is hard to break into simpler materials [4].

Water-soluble polymers prepared from acrylic acid, maleic anhydride, methacrylic acid, and various combinations of these monomers are not biodegradable. These water-soluble polymers are extensively used as detergent builders, scale inhibitors, flocculants, thickeners, emulsifiers, and paper-sizing agents. They are found in cleaning products, foods, toothpaste, shampoo, conditioners, skin lotions, and textiles. Hence, the toxicity in water bodies is increasingly alarming as these polymers are not biodegradable, potentially causing serious alterations to complex aquatic ecosystems. So, there is an urgent need for water-soluble biopolymers by modifying existing natural biopolymers such as starch and cellulose.

Carboxymethyl cellulose (CMC) is water soluble because it has different degrees of carboxymethyl substitution. Hydroxyethyl cellulose (HEC) is used as a thickener in drilling fluids and as a fluid-loss agent in cementing. CMC and HEC are polysaccharide-derived polymers. Higher levels of modification are required to attain a desired performance, but the rate of the extent of biodegradability decreases. Poly(vinylpyrrolidone) (PVP) is soluble in water and other polar solvents. It is hygroscopic in nature and forms films easily. Hence, PVP provides excellent wetting properties in making a coating or an additive to coatings. Pure PVP is edible and is used as a binder in pharmaceutical tablets, solutions, ointment, pessaries, liquid soaps, and surgical scrubs. PVP is thus extensively used because of its thickening and complexing property. PVP has polar moiety, which can be easily biodegradable by microorganisms. Poly(ethylene glycol) (PEG) is also a water-soluble biopolymer found in various applications such as toothpastes, as the separator, electrolyte solvent in lithium polymer cells, in phenol skin burns to deactivate any residual phenol as a dispersant, as a polar stationary phase for gas chromatography, as an anti-foaming agent, as lubricant eye drops, etc. PEGs are available in different molecular numbers (100–10,000), finding application in the medical, energy, and engineering fields. With the backbone of PEG being alkyl groups along with hydroxyl groups as the functional group, this makes it a biodegradable biopolymer.

Poly(vinyl alcohol) (PVA) is a water-soluble biodegradable biopolymer. Its dissolves partially at lower temperatures and dissolves rapidly at higher temperatures, making it almost stable at room temperature. Polyvinyl acetate is hydrolyzed in the presence of acids or alkalis to get PVA. PVA has a colloidal property and forms emulsions in an aqueous medium. The major use of PVA is in the textile industry as it brings about excellent resistance to abrasion as well as remarkable tenacity in textiles. It is also extensively used in energy devices as biopolymer electrolytes.

Natural cellulose biopolymer molecules have a molecular weight ranging from 300,000 to 500,000 Da. Cellulose has three hydroxyl groups that can be chemically modified based on most commercially important cellulosic polymers. The derivatization of cellulose mainly falls into two types: cellulose ethers and cellulose esters. Cellulose ethers find wide application in the food, pharmaceutical, paper, cosmetic, adhesive, detergent, and textile industries. Cellulose esters are prepared by either a fibrous or solution acetylation process. The fibrous acetylation process is less common due to its difficulty in isolation. The solution acetylation process is widely used commercially, but it requires a higher purity of cellulose. For this process the cellulose must contain a minimal amount of lignin and hemicelluloses impurities as well as alpha-cellulose content of at least 95%.

Chitosan is a natural linear polysaccharide composed of randomly distributed β-(1-4)-linked D-glucosamine (deacetylated unit) and *N*-acetyl-D-glucosamine (acetylated unit). Chitosan is obtained by deacetylation of chitin, which is found in the exoskeleton of crustaceans. Commercially, chitosan is derived from the shells of shrimp and other sea crustaceans. The amino group in chitosan has a p*K*a value of 6.5, hence making it dissolve only in an acidic medium. This has made chitosan useful in biomedical fields and water-purifying membranes. The film-forming property of chitosan is excellent and hence it is used as a biopolymer electrolyte in various energy devices.

Poly(styrenesulfonic acid) (PSSA) is a water-soluble biopolymer based on the polystyrene monomer. Usually, PSSA is prepared by polymerization or copolymerization of sodium styrene sulfonate or by sulfonation of polystyrene. PSSA, a polyion biopolymer, is used as a superplastifier in cement and in ion-exchange applications as well as a dye-improving agent for cotton and as proton exchange membranes in fuel cell applications. Therefore, PSSA is in the ionic/charged groups and when doped with conducting salts, this makes it a better biopolymer electrolyte to be used in energy devices. Table 1.1 shows the list of biopolymers and its sources from which they have be obtained.

1.1.2 Opportunity

Preparing biopolymers provides a great opportunity for developing green chemistry in industries. Deriving polymers from renewable sources such as annually renewable crops and agroindustrial waste streams instead of petroleum reserves will lead to a cleaner ecosystem. Biotechnology has certainly helped over the past decade in genetically modifying the metabolic pathways in microbes so that they can more efficiently convert inexpensive feedstocks (such as molasses,

TABLE 1.1 List of Biopolymers and Their Sources

Biopolymers	
Sources: Plant	
Lipid and phenolic biopolymers	Succinoglycan
Biopolymer cutin	Hyaluronan
Lignin-based	Agarose
Cellulose and its derivatives	Dextran
Alginate	Wood
Pectin	Gums (gum arabic, guar gum, gum tragacanth, gum karaya, and locust bean gum)
Fibrinogen	Isoprene biopolymers: natural rubber
Starch (wheat, potatoes, maize, etc.)	Cis and trans polyisoprene
Microorganism based biopolymers and genetically engineered biopolymers	
Microbial polysaccharides	Xanthan
e-Poly-L-lysine (e-PL)	Pullulan
Polyhydroxyalkanoates	Recombinant protein polymer
Poly(lactic acid)	Elastin
Poly(3-hydroxybutyrate) (PHB)	Collagen
Hyaluronan (polysaccharide)	Cellulose-derived biopolymers-based hydrogels
Bacterial cellulose	Supermacroporous cryogel matrix from biopolymers
Poly(3-hydroxybutyrateco-3-hydroxyvalerate) (PHBV)	Cryogel (protein)
Polyanhydrides	
Animal	
Chitin	Albumin (protein)
Chitosan	Silk
Casein	Whey
Heparin	Collagen
Leather (protein)	Gelatin
Keratin	

TABLE 1.1 List of Biopolymers and Their Sources—cont'd

Biopolymers	
Synthetic	
Polyesteramides	Poly(butylene succinate-*co*-adipate)
Polycaprolactone	Poly(butylene sebacate)
Aliphatic copolyester	Poly(glycolic acid)
Aromatic copolyester	Polyglycolide
Poly(vinyl alcohol)	Poly(propylene fumarate)
Poly(butylene adipate)	Polyphosphazenes
Poly(butylene succinate)	Poly(4-hydroxybutyrate).

starch, and waste lipids) to biopolymer building blocks [5]. Hence, genetic modification in plants will benefit renewable feedstocks and help in the cost-effective manufacturing of safe biopolymers. The ecological balance is also maintained because the biopolymers taken from nature will be returned to nature in a span of 1 year. Water treatment plants will boost the biodegradation of water-soluble biopolymers and this will reduce burring of biopolymers for soil biodegradation. If the technology and infrastructure grow in the biodegradation of these biopolymers, then we can treat the generated biowaste into valuable compost, chemical intermediates, and energy through aerobic and anaerobic processes. Nonetheless, these biopolymers having a low shelf life will continue to be indemand wherein products have relatively short-use lifetime. Therefore, the use of biopolymers in articles over a lifetime of years needs to be attended. The use of biopolymers in the field of energy storage is becoming popular nowadays because the shelf life of an energy device made from synthetic polymer is 3–4 years and are disposed to landfills without any treatment. Biopolymers as biopolymer electrolytes are shown to have the same shelf life during extensive use in the energy device under proper packing and have a relatively similar specific capacitance. After disposal, these biopolymer electrolytes are easily biodegradable in composts and other materials such as heavy metals and nondegradable materials can be recycled.

1.2 POLYMER ELECTROLYTES

Polymer electrolytes are being remarkably emphasized as a chemical science that provides polymers with new functionalities. Consequently, multidisciplinary research has emerged to further rationalize the process and bring about innovative materials necessary in key roles such as the ionic conductor, mechanical separator, and the flexible electronic insulator. Hence, polymer electrolytes are mainly used in electrochemical devices such as batteries, electrochromic devices, solar cells, and supercapacitors. Polymer electrolytes are potential materials for solving the never-ending demand for high energy density in energy devices. Polymer electrolytes are defined as linear macromolecular chains bearing a large number of charged or chargeable groups when dissolved in a suitable solvent.

Polymer molecules that have one or a few ionic groups, in most cases terminal and anionic, are called macroions. These are primarily living polymers wherein polymer molecules that are present in a polymerizing reaction system grow as long as monomers (e.g., esters or nitriles of methacrylic acid) are continuously supplied. The ionic charge of the macroion gets transferred to the next monomer added, and this process continues by keeping the macroion charged for the addition of further monomers. Polymers having a considerable number of ionic groups and a relatively nonpolar backbone are known as ionomers. Those polymers with a number of ionic groups and that can dissolve in water are known as polyelectrolytes. Polymers with a much higher number of ionic groups get cross-linked or undergo three-dimensional polymerization, which constitutes the technically important group of ion exchangers.

An important unresolved area in the field of polymer electrolytes concerns the role of the anion in ionic conductivity and the degree of ions present in some systems. It can be shown that the net cation mobility, in contrast to anion, is negligible. To act as a successful polymer host, a polymer or the active part of a copolymer should generally have a minimum of three essential characteristics:

- Atoms or groups of atoms with sufficient electron-donating power to form coordinate bonds with cations.
- Low barriers to bond rotation so that the segmental motion of the polymer chain can take place readily.
- A suitable distance between coordinating centers because the formation of multiple intrapolymer ion bonds appears to be important.

Wright's report, which talks about semicrystalline structure complexes between PEO and salt, led to a spurt in research to improve the conductivity of polymer electrolytes worldwide [6]. The correlation between the amorphous phase and ion conductivity led to curious attempts to study the electrical properties of this polymer electrolyte [7]. The first proposal of Armand [8] shows the use of a solid polymer electrolyte in lithium batteries. With modernization, the demand for light, high-performing, portable, and cost-effective energy devices also increases. This leads to increased interest in the study of the polymer electrolyte because it plays a vital role in fulfilling these needs. Polymer electrolytes must exhibit ionic conductivity in the range 10^{-3}–10^{-2} S cm^{-1} at room temperature. Electrochemical stability is an important parameter, but the instability brings about irreversible reactions leading to the gradual fading away of capacitance [9]. Even mechanical and thermal stability during charge/discharge cycles are crucial for the long durability of energy devices.

Uma et al. [10] have studied PMMA complexed with an Li_2SO_4 thin film polymer electrolyte prepared using the solution-casting technique. The complexation of salt with PMMA was confirmed by structural and thermal studies. The results of the ionic conductivity measurements in these electrolytes have also been reported.

Stephan et al. [11] have reported a novel polymer membrane comprising a poly(vinylidene difluoride hexafluoropropylene) (PVDF-HFP) copolymer prepared by the phase-inversion technique with two different nonsolvents. The membranes were gelled in an electrolyte solution of 1 M of $LiPF_6$ in EC/DMC and were subjected to A.C. impedance analysis at temperatures ranging from −30°C to 70°C.

The growth of this polymer electrolyte field over a diverse range of innovative modifications based on applications has gone through three main stages: solid polymer electrolyte systems, blend polymer electrolyte systems, and gel polymer electrolyte systems. These systems progressively satisfy the requirements for use in fuel cells, supercapacitors, secondary batteries, sensors, dye-sensitized solar cells, and microelectronic devices. Therefore, the polymer electrolyte stands as one of the keystones in the electronic storage and conversion areas in the current century.

1.3 BIOPOLYMER ELECTROLYTES

The development of new materials that can be applied as solid electrolytes has led to the creation of modern systems of energy generation and storage. Among different poly (ethylene oxide)-based electrolytes, natural polymers, such as hydroxyethylcellulose, hydroxypropylcellulose or carboxymethylcellulose (polysaccharides), starch, polyvinyl alcohol, polyethylene glycol, and chitosan or proteins such as gelatin are considered polymer electrolytes. Those polymers can undergo biodegradation as well as show the ionic conductivity improvement by inorganic dopants, acids, and gelation; these polymer electrolytes are called biopolymer electrolytes.

1.4 CLASSIFICATION OF BIOPOLYMER ELECTROLYTES

We have classified the biopolymer electrolytes based on their sources and origins, meaning that there are two types of biopolymer electrolytes: (a) synthetic and (b) natural. Some natural biopolymer electrolytes that are widely used in research are chitosan, starch, and cellulose materials. Biopolymer electrolytes can also be categorized into five different types based on their physical state and composition: (a) gel biopolymer electrolytes, (b) hydrogel biopolymer electrolytes, (c) solid biopolymer electrolytes, (d) blend biopolymer electrolytes, and (e) composite polymer electrolytes.

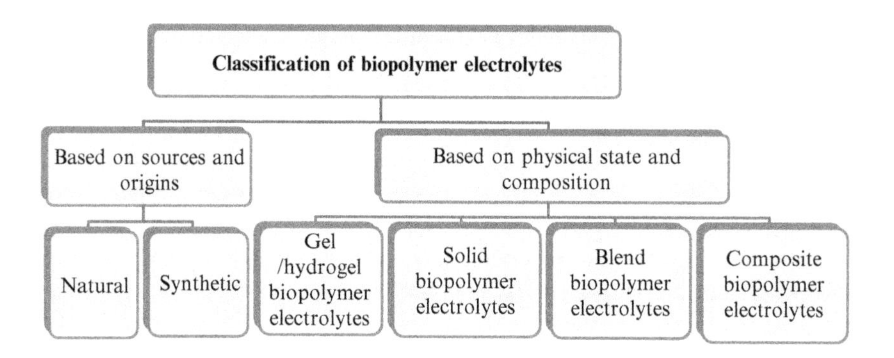

Biopolymers or biodegradable polymers are a newly emerging field. In recent years there has been an increase in interest in biopolymers. The applications of biopolymers earlier were focused on three major areas: medical, agricultural, and consumer goods packaging. The storage, production, and distribution of energy are among the main concerns of modern industry and society. Space development, the creation of new types of memory, and new computer architecture along with biomedical devices and microsensors are all areas that could benefit from the development of solid-state ionic conductors.

The use of portable electronic devices such as mobile phones, laptops, and power banks has been increasing, promoting interest in biopolymer research into their power sources. These devices contain power like supercapacitors usually configured with a liquid electrolyte between the two electrodes. The key component of the supercapacitor is the electrolyte. The appropriate choice of this component is ruled by a series of requirements, which include high ionic conductivity, good mechanical properties, and compatibility with electrode materials [12]. The common problems in the liquid electrolyte-based supercapacitor are the leakage of flammable electrolyte, the production of gases upon overcharge or overdischarge, and thermal runaway reaction. Therefore, the application of solid-state electrolyte devices can solve these problems. The solid-based polymer electrolyte supercapacitors are estimated to surpass the performance of conventional liquid electrolyte systems. Nonetheless, little attention has been paid to the polymer electrolytes that in large amounts are also hazardous to the environment as these are not biodegradable or recyclable. Intensive work is also being done on developing and using environmentally conscious materials, or ecomaterials, during manufacturing, which can help reduce the environmental impact of many products throughout all phases of the product's life cycle. Based on the source or origin, the biopolymers are generally from renewable resources that can be categorized into three groups: (a) natural polymers such as polysaccharides or proteins; (b) bioderived monomers used for synthetic polymers such as poly (lactic acid); and (c) biopolymers from microbial fermentation such as polyhydroxybutyrate, hyaluronic acid, poly-γ-glutamic acid, etc.

1.5 DOPANTS

1.5.1 Lithium Salts as Dopants in Biopolymer Electrolytes

The desired properties of lithium salt for use in energy applications must be as follows:

- Lithium salts have to completely dissolve in the applied solvent at the optimized concentration and ions should be able to transfer through the biopolymer matrix.
- Lithium ions should not undergo oxidative decomposition at the cathode.
- Lithium ions must be inert to the electrolyte solvent/biopolymer.
- Both anion and cation should be inert toward other cell components.
- The anion must be nontoxic and remain thermally stable at the energy device's working conditions.

$$\text{Salt}\begin{pmatrix} LiClO_4, LiPF_6, LiBF_4, LiAsF_6, LiCF_3SO_3, LiN(CF_3SO_2)_2, \\ LiC(CF_3SO_2)_2, Li^+[CF_3SO_2NSO_2CF_3]^-(LiTFSI), etc. \end{pmatrix}$$

The existence of polar groups in biopolymers is necessary to dissolve salts and form stable ion-biopolymer complexes. The lattice energy may be compensated by factors such as a low value of cohesive energy density and vacancy formation, favored by a low glass transition temperature, Lewis acid-base interactions between the coordinating sites on the polymer and the ions, and long-range electrostatic forces such as cation-anion interaction energies.

The role of the anion and cation are important in determining the transference number and ionic conductivity. The hard acid cation Mg^{2+} is expected to form very stable complexes with oxygen as the electron donor in the polymer, whereas Hg^{2+} (soft acid) shows only weak interaction. When the transference number is measured, the Mg^{2+} ions were immobile and the Hg^{2+} ions were mobile. So promoting complex formation may also have consequential effects on cation mobility. Most anions are destabilized on passing from the polar protic medium through to a less polar one, the destabilization being greatest when the charge density and basicity of the ion is low.

1.5.2 Acids as Dopants in Biopolymer Electrolytes

The advantage of proton-conducting biopolymer gel systems consisting of H_2SO_4, o-H_3PO_4, and HCl in comparison with lithium salt-doped polymer electrolytes arises from their potentially higher conductivity and smaller ionic radius. Furthermore, with biopolymer complexes with low inorganic salt concentrations, ion pairing is possible, while at high concentration the formation of large ionic aggregation may occur. Thus, lithium salts containing polymer electrolytes lack higher power density and mainly depend upon electron-rich functional groups of polymers for higher conductivity. Even EDLCs having these polymer electrolytes exhibited higher capacitance than those with organic electrolytes, Ils, and lithium salt-doped polymer electrolytes. Ambient temperature conductivities obtained for some of these proton-conducting biopolymeric electrolytes were higher than $10^{-3}\,S\,cm^{-1}$ [13,14]. There are serious limitations regarding the availability of fast proton solid conductors, especially at ambient temperatures. However, there is very little in the available literature in which strong acid-doped biodegradable polymer electrolytes were used in an EDLC application. The disability of complete dissociation is a major problem of using weak acids [15]. Strong acids are known to degrade the polymer system even under the operating potential of 1V; this creates an opportunity for preparing stabilized acid-doped polymer systems and makes their use in EDLCs more demanding.

H_3PO_4 has proved to be a proton conductor in poly(ethylenimine) [16], poly(vinyl alcohol) [17], and poly(silamine) [18] chitosan/iota-carrageenan [19]. The source of proton conduction can be obtained by self-ionization and self-dehydration of H_3PO_4 [20]. H_3PO_4 also acts as a plasticizer [21]. The addition of H_3PO_4 may further induce an esterification reaction with hydroxyl groups of cellulose derivatives, but acid hydrolysis of β-glucoside bonds would be minimized in the presence of borax.

1.5.3 Alkaline Dopants in Polymer Electrolytes

Alkaline dopants are generally the salts containing OH^- ions. Several theories along with experimental results have shown the involvement of OH^- ions in the conduction via hypercoordinated or per-solvated complexes. When approached by another H_2O molecule, the hypercoordinated $OH^-(H_2O)_4$ complex gets activated, wherein the OH^- forms a hydrogen bond with the water molecule. The proton from the water molecule gets transferred to the hypercoordinated complex ion and the new hypercoordinated complex carrying the negative charge is formed. When another water molecule approaches this complex again, the proton gets transferred and hence the negative charge continues over solvent molecules. Similar to proton hopping, this mechanism is also referred to as OH^- hopping, but this ionic conduction is slower than proton conduction. Alkaline dopants such as KOH, tetraalkylammonium hydroxide (TAAOH) are already incorporated in supercapacitor studies. Alkaline-based polymer electrolytes have not been explored much other than KOH, hence a tremendous scope lies ahead in prepared alkali-doped biopolymer electrolytes.

1.5.4 Plasticizing Salts/Ionic Liquids

Ionic liquids are low-temperature molten salts. They consist of a large cation and a charge-delocalized anion. These ionic liquids are similar to organic electrolytes and can have vapor pressure up to ~250–450°C. The presence of organic ions provides unlimited structural variations.

The following properties of ionic liquids are advantageous for energy devices: thermal and chemical stability, negligible volatility, flame retardancy, moderate viscosity, high polarity, low melting point, high ionic conductivity, and solubility (affinity) with many compounds.

Plasticizers are used to reduce the crystalline phase of polymer and increase the ambient temperature ionic conductivity in polymer-salt complexes. Nevertheless, being low molecular weight organic liquids, they are volatile. If the plasticizer is improperly blended with the polymer, even slow evaporation can lead to the disintegration of polymer electrolyte films. Also, the addition of some of the plasticizers such as ethylene carbonate or propylene carbonate in large quantities leads to a loss of dimensional stability of the polymer electrolyte films. Therefore, the uses of plasticizing salts, which are nonvolatile ionic compounds with bulky anions, are relatively advantageous. A few of the salts mixed for the desired property are imidazolium, pyrrolidinium, and quaternary ammonium salts as cations and bis(trifluoromethanesulphonyl)imide, bis(fluorosulphonyl)imide, and hexafluorophosphate as anions. Other commonly used plasticizing salt combinations are lithium bis(trifluoromethane sulfonyl)imide (LiTFSI), $LiN(SO_2CF_3)_2$, lithium tris(trifluoromethane sulfonyl) methide (Li Tri TFSM), Li $C(SO_2CF_3)_3$, lithium pentafluoro sulfur difluoromethylene sulfonate $LiSO_3CF_2SF_5$, and lithium bis(trifluoromethane sulfonyl)methane $LiCH(SO_2CF_3)_2$.

1.6 SOLID BIOPOLYMER ELECTROLYTES (SBPE)

Solid biopolymer electrolytes are a liquid-free, high molecular weight, polar polymer host having an ionically conducting phase formed by dissolving salts. Many polymer electrolyte materials will exhibit to a greater or lesser extent the following properties:

- Adequate conductivity for practical purposes.
- Good mechanical properties.
- Chemical, electrochemical, and photochemical stability.
- No possibility of leakage.
- Ease of processing.
- Shape flexibility.
- Lowering the cell weight-nonvolatile, all-solid-state cells do not need a heavy steel casing.

For the success of these properties, low interfacial resistances, a high ionic transport number, and excellent conductivity are required. This result shows the strong tendency of polymers to crystallize, and crystalline phases are characterized by much lower conductivities than the amorphous polymers.

In general, the polyelectrolyte solutions contain a single species of polymer and one species of counter ions only. Eventually, the solution may also contain a single species of low-molar-mass electrolytes (to be called the "doped salt," but it may be a strong acid or a base) having common counter ions with the polyelectrolyte unless otherwise specified and assumed not to interact chemically with the polyelectrolyte.

1.6.1 Polymer Dissolution

Solutions of polyelectrolytes exhibit a behavior that may differ considerably from that of uncharged macromolecules of low-molar-mass electrolytes. The origin of this specificity lies in the combination of properties derived from those of long-chain molecules with properties that result from charge interactions. This combination is not a simple superposition as there is a mutual influence of the characteristics of both types of properties. Thus, the presence of charges on the chain leads to intra-and inter macromolecular interactions, which may be stronger and of much longer range than in the case of uncharged macromolecules. This may have a strong influence on both the thermodynamic and dynamic properties of polyelectrolyte solutions, particularly if in the solution, the electrostatic potential arising from the charges fixed on the macromolecular chains is not sufficiently screened. Uncharged biopolymers tend to coil up in a dissolved state, but polyelectrolytes having charged groups have repulsive forces and stretch out with violent chain movement, which eventually breaks up. Hence, accumulation of like charges closely spaced along the polymer chain leads to a certain rigidity of the polyelectrolytes. Therefore, an unexpected increase in viscosity is observed in these polyelectrolytes.

The strength of the electrostatic interactions may be moderated by increasing the concentration of added salt, which results in a screening effect by the small ions; however, high concentrations of salt may also affect the solvent quality. These interactions strongly affect not only the average dimensions of the polyelectrolyte chain, but also the intra macromolecular dynamics of the chains. The screening effects are primarily the result of the interactions between the bound and mobile charges (i.e., between the charged chains and the small ions in solution). These interactions are not completely comparable to the charge-charge interactions occurring in solutions of low-molar-mass electrolytes. This occurs from the divergent behavior of the bound and mobile charges, even in dilute polyelectrolyte solutions. The former is considerably restricted in its motions by the chain on which it is fixed. Local charge fluctuations arising from charges bound to the same chains will therefore be highly correlated and hence lead to a fundamental asymmetry between the bound and mobile charges. On the molecular level, charges bound to a same chain will have to cluster to some extent around each other in all possible configurations, in contrast to the mobile charges, which, in principle, can move independently through the entire volume of the solution. Around the chain-bound charge clusters, the electrostatic potential will be much higher than elsewhere in the system, but this potential may fluctuate in accordance with the conformational fluctuations of the macromolecular chain.

1.6.2 Movements of Ions in SPE

After the dissolution of salt in the polymer solution, the polymer must solvate the ions by overcoming the lattice energy of the ionic salt and thereby form a complex [22]. The three main criteria in forming a complex are:

(a) Electron pair donicity (DN)
(b) Acceptor number (AN)
(c) Entropy term

The efficiency of a polymer to solvate the ions is given by the term DN, which follows the Lewis acid base concept. Therefore, for the polymer to act as a host polymer electrolyte, it should have donor sites such as oxygen, sulfur, or nitrogen, either in the backbone or in a group attached in the form of a side chain to the polymer. Correspondingly, the solvation of the anion is described by the AN term, which is Lewis base. The DN number must be greater than AN. Because the PEO has a higher DN than cations, it will effectively solvate cations possessing counter anions that are bulky delocalized anions such as I^-, ClO_4^-, BF_4^-, or $CF_3SO_3^-$. The entropy term describes the spatial movement of the polymer in the solvating unit. In the PEO containing ethylene, oxy (CH_2CH_2O) have the most favorable spatial orientation of the solvating units. Smaller ions such as Li^+ are easily solvated and form polymer-salt complexes. In order to get solvated by poly(ethylene phthalate) (PEP) by larger cations such as Na^+, K^+ etc., there is a necessity for bulky counter anions such as I^-, SCN^-, or $CF_3SO_3^-$.

Along with the above factors, the polymer must lack extensive intermolecular hydrogen bonding because it will affect the solvation ability of the polymer. The other factor is that it should withstand high torsion by keeping itself flexible. This can be achieved by using a polymer having a low glass transition temperature T_g. Low T_g favors a large segmental motion of the polymer chain and thus increases the ionic conductivity. Ion transport relies on local relaxation processes in the polymer chains that may provide liquid-like degrees of freedom, giving the polymer properties similar to those of a molecular liquid. The macroscopic properties that are similar to those of a solid are the result of chain entanglements and possibly cross-linking.

The transportation of ions in biopolymer electrolytes is considered to take place by a combination of ion motions between ion coordinating sites. Ratner et al. [23] have developed a dynamic percolation model for description of ion transport in polymer electrolytes. This macroscopic model characterizers the ionic motion in terms of hoping between neighboring positions. The free anions and cations take time to coordinate with the polymer host and this linkage changes with time. Consider M^+ = metal ion and B = Lewis base site on the polymer, a single linkage M^+-B is expected to change in time. The "hop" of a cation then corresponds to a completed exchange of one ligand. This process is sketched in Fig. 1.1. Fig. 1.1A shows the motion of ions coupled to that of the polymer chain; lateral displacement is brought about by 180-degrees bond

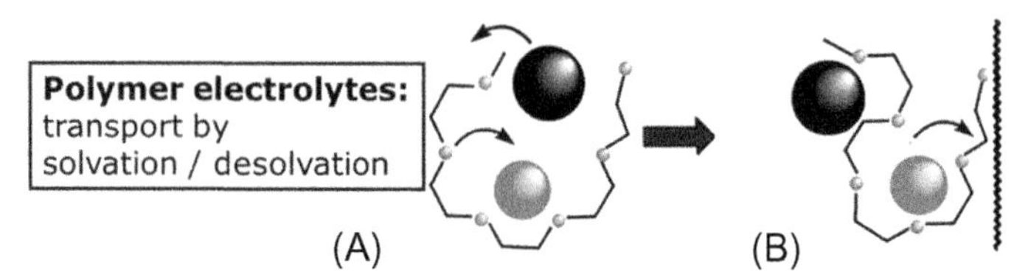

FIG. 1.1 Cation transport mechanism in a polyethylene oxide-based polymer electrolyte. (A) Solvation; (B) Desolvation.

rotation at the C—O bond. Fig. 1.1B presents the first step in the transfer of a cation between chains. Anions may also be involved as part of either an ion pair or an ion triplet.

If the salt concentration is increased, ion clusters are formed. The mechanism of charge transfer between the ionic clusters now competes with the segmental motion of the biopolymer. Charge transfer becomes relatively faster with increased ionic clusters as [24]

$$M^+X^-M^+ + X^-M^+ \rightarrow M^+X^- + M^+X^-M^+$$

These activated cation jumps would not only be limited to such small ionic clusters or ion pairs but would also involve cation-biopolymer bond exchange.

1.6.3 Proton Conduction Mechanisms

When doped with suitable acids, the biopolymer gets dispersed in the matrix. Based on the interaction of protons with biopolymer chains, the proton transfer phenomena in biopolymer electrolytes follow two principal mechanisms: the chemical mechanism (vehicle type) and the Grotthus mechanism [25]. In media that support strong hydrogen bonding, the Grotthus mechanism is preferred; the vehicle mechanism is characteristic of a species with weaker bonding. When the temperature is increased, the Grotthus mechanisms are progressively dominated by vehicle mechanisms as the fast motion of chains breaks the hydrogen bonds between the proton and the chains.

In the proton-doped biopolymer electrolyte, the proton remains shielded by the electron density along its entire diffusion path; hence the motion of free protons inside the matrix is never seen. Nonetheless, translational transportation of protons as a bigger species is observed in rare cases; this is the vehicle mechanism. In this mechanism, the proton diffuses through the medium together with a vehicle (for example, with H_2O as H_3O^+). The counter diffusion of deprotonated vehicles (H_2O) allows the net transport of protons. The observed conductivity, therefore, is directly dependent on the rate of vehicle diffusion. If the vehicles show pronounced local dynamics, but reside on their sites, the protons are transferred by hopping from one vehicle to the other by hydrogen bonds. Reorganization of water molecules along with the biopolymer simultaneously

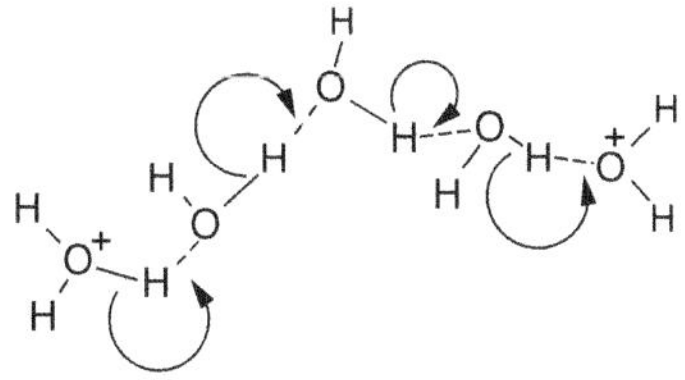

FIG. 1.2 Proton transportation by Grotthus mechanism.

takes place, which leads to the building of an uninterrupted path for proton migration (i.e., proton diffusion pathway). This mechanism is known as the Grotthus mechanism. Therefore, this mechanism depends mainly on the rates of proton transfer Γ_{trans} and reorganization of its environment Γ_{rep} and all these rates are directly connected to the diffusion of protons ($\Gamma_D, \Gamma_{trans}, \Gamma_{rep}$) (Fig. 1.2).

1.6.4 Dependence of Cation Mobility on the Relative Molar Mass of the Polymer Host

High molecular mass biopolymer electrolytes doped with cations having a low solvent exchange rate will not exhibit significant long-range motion. Hence, drawing current from these biopolymer electrolytes is not possible. For lower molecular mass biopolymer electrolytes, mechanisms involving polymer chain diffusion become important. For these dissociative steps, involving the ion-solvent bond scission is not necessary. Simple linear polymers of comparatively low molar mass have chain diffusion coefficients that are inversely proportional to their molar masses. The Rouse-Zimm model, originally proposed to describe the viscoelastic properties of dilute solutions of coiling polymers, is also found to describe successfully the motion of such molecules in the pure liquid state, correctly predicting the molecular weight dependence of the self-diffusion coefficient [24]. Above a critical value of the molar mass, chain entanglement becomes significant and a new model of polymer dynamics must be considered. At short times, a high molecular weight amorphous linear polymer behaves as a rubber because the "knots" formed by two polymer strands do not have time to become untied. At longer times, Brownian motion results in local disentanglement, allowing the chains to slide past one another.

The solid biopolymer as electrolyte has been constantly studied in the field of energy devices with different dopants and modes of preparation; here is a brief review on solid biopolymer electrolytes. Rodriguez et al. [26] have prepared SPE films consisting of mixtures of poly(vinylpyrrolidone) and $LiClO_4$ with various mass ratios using the dip-coating method. A conductivity of $3.3 \times 10^{-3}\,S\,cm^{-1}$ at 60°C was obtained; it was reported that a residual amount of solvent was important in preserving the ionic conductivity. They demonstrated that these films are used as transparent SPE in supercapacitors.

1.7 BLEND BIOPOLYMER ELECTROLYTES (BBPE)

1.7.1 Introduction of BBPE

BBPEs are a physical mixture of two or more polymer chains forming a homogeneous (liquid) solvent-free system. Sometimes, though the various phases are chemically bonded together wherein the ionically conducting phase is formed by dissolving inorganic salts. Hydrogen bonding, charge transfer interactions, and dipole-dipole forces form the basis for the miscibility of polymer blends. The manifestation of properties of polymer blends depends upon the miscibility of the components and structure. Prior to the preparation of a polymer blend electrolyte, one must know the polymer blend interaction studies.

Blended mixtures may offer distinct properties wherein the property of each polymer enhances the overall property of the mixture. The property mixing of polymeric blends is dependent on a number of factors, one of the major ones being the miscibility of the polymers in one another. In some polymer blends, a change in the ratio of mixing two polymers can bring about miscibility. Here, in a microscopic scale, polymer blends will be divided into homogeneous or heterogeneous polymer blends [27].

Based on the fact that the polymers are compatible or miscible, the following blends are listed:

(a) *Compatible blend*: A mixture of polymer in which the overall property is enhanced when compared with its nonblend components. Herein, there are only physical forces of attraction that keep the polymers in single phase.

(b) *Incompatible blend*: A mixture of polymers in which the overall property is less when compared with its nonblend components.

(c) *Polymer alloys*: Compatibilizers are used to improve the property balance; these blends are commercial polymer blends having interface.

(d) *Miscible blend*: A mixture of polymers in which a homogenous phase exists at a microscopic level due to physical forces and hydrogen bonding.

(e) *Immiscible blend:* A mixture of polymers in which a heterogeneous phase exists at a molecular level. The compositions of the separated phases are identical to the pure polymer components prior to blending.

(f) *Partially miscible blend:* A mixture of polymers in which a heterogeneous phase exists at a molecular level. The compositions of the separated phases are not identical to the pure components prior to blending.

(g) *Interpenetrating polymer network (IPN)*: One of the polymers is synthesized or cross-linked in the presence of another polymer, forming a network. A compatible blend that has commercial possibilities may be immiscible, and a miscible polymer blend may lack commercial applications due to other factors such as cost, source of raw materials, safety, and environmental issues such as recyclability.

1.7.1.1 Preparation of Polymer Blends

- Solution mixing—laboratory or paint industry, but the problem is removing a solvent.
- Interpenetrating networks—crosslinked materials.
- Melt mixing (physical, reactive)—most important method in industry.

When two different polymers are dissolved effectively in a common solvent, there is a fast formation of thermodynamic equilibrium among the polymers at the molecular level [28]. The difficulty with this procedure is due to the fact that many polymers become incompatible above a certain concentration when their solutions in a common solvent are mixed. Hence, polymers may not be miscible when they are mixed together, but have dissolved in a common solvent. Even the way they have been made to form a film also matters because a method such as evaporation is a slow process that can lead to inhomogeneities. During fast precipitating processes, this is far less the case, for example, by addition of a precipitating agent during spray precipitation or by rapid evaporation.

1.7.2 Miscibility and Thermodynamic Relationships of Biopolymer Blends

In the preparation of a blend polymer, the most important characteristic that must be known is the phase behavior in order to interpret the mechanism of ions. The miscible blend polymer electrolytes could provide a homogeneous pathway for the conduction of the ion rather than specified to a particular polymer in an immiscible blend system. Molecular weight is one of the lead factors in bringing about miscibility. Mostly in low molecular weight polymers, the combinatorial entropy contribution is higher than high molecular weight polymers. This criterion makes solvent-solvent mixtures provide a broader range of miscibility than polymer-solvent combinations. Furthermore, in polymer-polymer mixtures, the range of miscible combinations is even smaller.

In order to explain the relationship relating to mixtures of two dissimilar components, the thermodynamic aspects illustrate the relationship as

$$\Delta G_m = \Delta H_m - T\Delta S_m \tag{1.1}$$

where ΔG_m is free energy of mixing, ΔH_m is the enthalpy of mixing (heat of mixing), and ΔS_m is the entropy of mixing. For miscibility to occur, ΔG_m must be smaller than 0. So, with an increase in temperature, that is, an increase in $T\Delta S_m$, the ΔG_m attain more negative values and components are miscible. Nevertheless, for high molecular weight polymers, $T\Delta S_m$ is less and ΔH_m increases, which leads to phase separation, that is, miscibility decreases with increasing temperature. The enthalpy of mixing can be solved using the solubility parameter concept used by Hildebrand:

$$\frac{\Delta H_m}{V} = (\delta_1 - \delta_2)^2 \phi_1 \phi_2 \tag{1.2}$$

where δ_1 and δ_2 are solubility parameter for components 1 and 2, the volume $(V) = V_1 N_1 + V_2 N_2$ and volume fractions ϕ_1 and ϕ_2 are represented by the equations

$$\phi_1 = \frac{V_1 N_1}{V_1 N_1 + V_2 N_2} \text{ and } \phi_2 = \frac{V_2 N_2}{V_1 N_1 + V_2 N_2} \tag{1.3}$$

The viscosity method is useful to study the miscibility of the polymer blends. The intrinsic viscosity of polymer solutions depends on the nature of the polymer, solvent, and temperature. It also depends on the polymer-solvent interaction. Hence, using the data obtained from viscosity, one can study the miscibility using thermodynamic relationships.

In order to determine the miscibility of the polymer blend solution, a simple and accurate measurement is using viscosity studies at different temperatures. The viscosity η of a liquid may be defined as the force per unit area required to maintain a unit velocity gradient between two parallel plates kept at a constant distance apart:

$$\eta = \frac{\frac{\text{Force}}{\text{Area}}}{\frac{\text{Difference in velocity}}{\text{Distance between plates}}}, (\text{Pa s}) \tag{1.4}$$

The viscosity may also be defined as the kinetic energy transformed into heat, per unit volume of liquid, divided by the square of the velocity gradient:

$$\eta = \frac{\text{Energy transformed/time} \times \text{volume}}{(\text{Difference in velocity/distance between plates})^2}, (\text{Pas}) \tag{1.5}$$

The fluidity Φ is the reciprocal of the viscosity:

$$\Phi = \frac{1}{\eta}, \left(\text{cm s g}^{-1}\right) \tag{1.6}$$

An apparatus used for determination of the coefficient of viscosity of the solution is an Ubbelhode-type capillary viscometer that allows reproduction of the flow times with an accuracy of 0.03 s. The viscometer consists of two limbs with bulbs and marks. Subsequent concentrations can be achieved by adding known volumes of pure solvent and mixing inside the viscometer itself. This yields data for computing flow times at different concentrations. The time of flow from one mark to the other both for water and solution are noted accurately at a given temperature. Then, the coefficient of viscosity of the solution is obtained using the formula

$$\eta = \frac{\rho \times t}{\rho_0 t_0 \eta_0}, (\text{Pas}) \tag{1.7}$$

where η is the shear viscosity (solution viscosity), ρ is the density, and t is the capillary flow time measure for the solution. η_0, ρ_0, and t_0 are the values corresponding to the solvent. From that viscosity the following terms are calculated.

$$[\eta] = \left(\frac{\eta - \eta_0}{\eta_0 C}\right)_{c=0} = \left(\frac{\ln(\eta/\eta_0)}{C}\right)_{C=0} \tag{1.8}$$

With the concentrations in grams of solute per 100 mL, this limit is known as the intrinsic viscosity. With concentrations in grams per milliliter it has been designated as the "limiting viscosity number" or, more recently, Flory 1953 as the Staudinger index [6].

$$\text{Relative viscosity} = \frac{\eta}{\eta_0} = \frac{t}{t_0} = \eta_r, (\text{dimensionless}) \tag{1.9}$$

$$\text{Specific viscosity} = \frac{\eta - \eta_0}{\eta_0} = \frac{(t - t_0)}{t_0} = \eta_r - 1 = \eta_{sp}, (\text{dimensionless}) \tag{1.10}$$

$$\text{Reduced viscosity} = \frac{\eta_{sp}}{c} = \eta_{red}, \left(\text{dL g}^{-1}\right) \tag{1.11}$$

$$\text{Inherent viscosity} = \frac{\ln \eta_r}{C} = \eta_{inh}, \left(\text{dL g}^{-1}\right) \tag{1.12}$$

$$\text{Intrinsic viscosity} = \left(\frac{\eta_{sp}}{C}\right)_{c-0} = \left(\frac{\ln \eta_r}{C}\right)_{c-0} = (\eta), \left(\text{dL g}^{-1}\right) \tag{1.13}$$

As the name is "relative viscosity," it is self-explanatory. Specific viscosity conveys the increased viscosity because of the presence of the polymer solution. Normalizing η_{sp} to concentration gives η_{sp}/c, which conveys the capacity of a polymer to cause the solution viscosity to increase per unit concentration of polymer. For nonideal viscosity behavior of polymer solutions, the η_{sp}/c will depend on c. The extrapolated value of η_{sp}/c at zero concentration is known as the intrinsic viscosity $[\eta]$. $[\eta]$ will be shown to be a unique function of molecular weight (for a given polymer-solvent pair) and measurements of $[\eta]$ can be used to measure molecular weight. Similar to η_{sp}, the inherent viscosity $\ln\eta_r$ is zero for pure solvent. With an increase in concentration due to the presence of polymer in the solution, $\ln\eta_r$ also increases. Normalizing $\ln\eta_r$ to concentration or $\ln\eta_r/c$ gives the inherent viscosity. In the limit of zero concentration, η_i extrapolates the same as η_{sp}/c and becomes equal to the intrinsic viscosity.

Viscometric method is frequently employed in polymer characterization for the determination of molecular weight and distribution, miscibility, branching degree, tacticity, crystallinity, etc. The presence of polymer molecules in particular solvents can give rise to a dramatic increase in viscosity, which is very much greater than that found for equivalent concentrations of low-molar-mass solutes. This is because of the enormous difference in dimensions between the polymer and solvent molecules; in good solvents the polymer coils are expanded even further. In general, the increase in viscosity depends upon a number of factors:

1. The nature of the solvent
2. The type of polymer
3. The molar mass of the polymer
4. The concentration of the polymer
5. The temperature

The shear viscosity exhibited by a dilute solution of long-chain flexible polymers results from the frictional resistance opposing the flow of solvent molecules past polymer segments and the resistance to perturbation of the configurational equilibrium of the polymer. Further, the bulk viscosity is due to the resistance of the polymer molecules against changing the volume they occupy in the solvent in response to a pressure change. It is dependent on solvent-polymer interactions that may not affect the shear viscosity. Shear viscosity of polymer solutions significantly contributes to the mechanical relaxation of solutions. At the molecular level, the viscosity of a polymer solution is a direct measure of the hydrodynamic volume of the polymer molecules. Hydrodynamic volume is the apparent volume occupied by the expanded or swollen molecular coil along with the imbibed solvent. It can be defined in terms of the expansion factor and the unperturbed end-to-end distance depends on the polymer molecular weight. It therefore follows that a polymer will exhibit a higher viscosity in a good solvent than in a poor solvent and that, in the same solvent; the viscosity will be directly proportional to the molecular weight.

1.7.3 Interaction Parameter (χ)

Thermodynamic properties of polymer solutions such as miscibility, swelling equilibria, and the colligative properties can be expressed in terms of the polymer-solvent interaction parameter χ. This unitless quantity was originally introduced by P.J. Flory and M.L. Huggins as an exchange interaction parameter in their lattice model of polymer solutions. In their definition, the quantity $kT\chi$ (k is the Boltzmann constant; T, the absolute temperature) is the average change in energy when a solvent molecule is transferred from pure solvent to pure, amorphous polymer. The following is the theory developed for the first time to arrive at the χ using the Gibbs free energy obtained from ultrasonic relaxation time data as well as from viscosity data. The properties of the polymer solution depend both on the nature and size of the polymer chains and also on the interactions between polymer and solvent molecules. The thermodynamic property that is used to monitor behavior of the system is Gibbs free energy (G). Consider the polymer solution containing n_1 solvent molecules and n_2 polymer molecules made up of x segments. The free energy G_{12} of the mixture of (n_1+xn_2) molecules is given by the relation,

$$G_{12}=n_1G_1+xn_2G_2 \tag{1.14}$$

where G_1 and G_2 are the respective free energies of the solvent and the polymer per unit mole, the free energy per unit mole of the solution ΔG_m (kJ mol^{-1}) en will be,

$$\Delta G_m=\frac{G_{12}}{n_1+xn_2} \tag{1.15}$$

$$\Delta G_m=\frac{n_1G_1+xn_2G_2}{n_1+xn_2} \tag{1.16}$$

$$\Delta G_m=\Phi_1G_1+\Phi_2G_2 \tag{1.17}$$

where $\Phi_1=\frac{n_1}{n_1+xn_2}$ and $\Phi_2=\frac{xn_2}{n_1+xn_2}$ are the volume fractions of the solvent and the polymer. The above equation is valid for an ideal case where there is no interaction; when there is interaction the free energy per unit mole of the solution will not be equal to $(\Phi_1G_1+\Phi_2G_2)$ and an additional term that specifies the interaction has to be included in the estimation of the free energy per unit mole of the solution. To compute this type of interaction in an analogy with the Flory-Huggins theory, the term $\chi\ n_1\Phi_2$ has been introduced as the additional term. Thus,

$$\Delta G_m=G_1\Phi_1+G_2\Phi_2+\chi n_1\Phi_2 \tag{1.18}$$

Here, all the energy components are measured in RT units. In this equation χ denotes the polymer-solvent interaction parameter and is also known as the Flory-Huggins parameter. It is to be noted that this term ΔG_m is different from the change in Gibbs free energy. In the case of dilute polymer solutions, the contribution of the term $G_2\Phi_2$ is taken to be small as Φ_2 is small compared to Φ_1. Hence, the above equation reduces to

$$\Delta G_m=G_1\Phi_1+\chi n_1\Phi_2 \tag{1.19}$$

Knowing ΔG_m and G_1, one can compute χ from the above equation. Thus

$$\chi=\frac{(\Delta G_m-G_1\Phi_1)}{n_1\Phi_2},\ \text{(dimensionless)} \tag{1.20}$$

To estimate χ, G_{12} is computed using Eqs. (1.16), (1.17). Knowing G_{12}, ΔG_m is calculated using Eq. (1.19). The Gibbs free energy of the solvent is calculated through both routes using the same equations and by substituting the corresponding data of the solvent. By dividing this obtained value by n_1, the Gibbs free energy per unit mole of the solvent (G_1) is obtained.

A linear dependence of the interaction parameter on the volume fractions Φ_2 of the solute can be written as the first approximation:

$$\chi = \chi_{o+}\sigma \tag{1.21}$$

where χ_o is the interaction parameter at infinite dilution and σ denotes the change in the interaction parameter per unit volume fraction of the polymer. This represents the slope in the plot of χ versus Φ_2. Further, the equation is valid only at low concentrations of the polymer solution.

1.7.3.1 Polymer-Polymer and Polymer Blend-Solvent Interactions

The interaction parameters between polymer-polymer and polymer blend-solvent are a measure of miscibility. The polymer-polymer interaction parameters (χ_i) have been computed from the Flory-Huggins theory with

$$\chi_i = \left(\frac{V_i}{RT}\right)(\delta_2 - \delta_1)^2 \tag{1.22}$$

where δ_1 and δ_2 are the solubility parameters of component polymers, respectively, and V_i, R, and T are the molar volume of the solvent, universal gas constant, and temperature (K), respectively. The blend-solvent interaction parameters have also been calculated according to the method adopted by Singh and Singh. The solubility parameters of the blend (δ) were calculated from the additively relationship,

$$\delta = X_1\delta_1 + X_2\delta_2 \tag{1.23}$$

where X_1 and X_2 are the mass fractions and δ_1 and δ_2 are the solubility parameters of the component polymers in the blend system. If the data observed for the net polymer-polymer interactions are higher than those observed for *blend*-solvent interactions for studied blend compositions at different temperatures means these polymers are compatible in nature.

1.8 GEL BIOPOLYMER ELECTROLYTES (GBPE)

1.8.1 Introduction of GBPE

GBPE is often known as a plasticized polymer electrolyte that is neither liquid nor solid, or conversely both liquid and solid. Gel contains a solid skeleton of polymers or long-chain molecules cross-linked intramolecularly or intermolecularly, entrapping an uninterrupted liquid phase. The chemical composition and other factors such as hydrogen bonding vary the chemistry of gels from viscous fluid to moderately rigid solids. Nevertheless, they are very soft and stretchy or "jelly like." Due to a large number of liquids filled in microspores most gel materials exhibit liquid like characteristics microscopically, but macroscopically have solid character. The presence of the ultraporous structure in

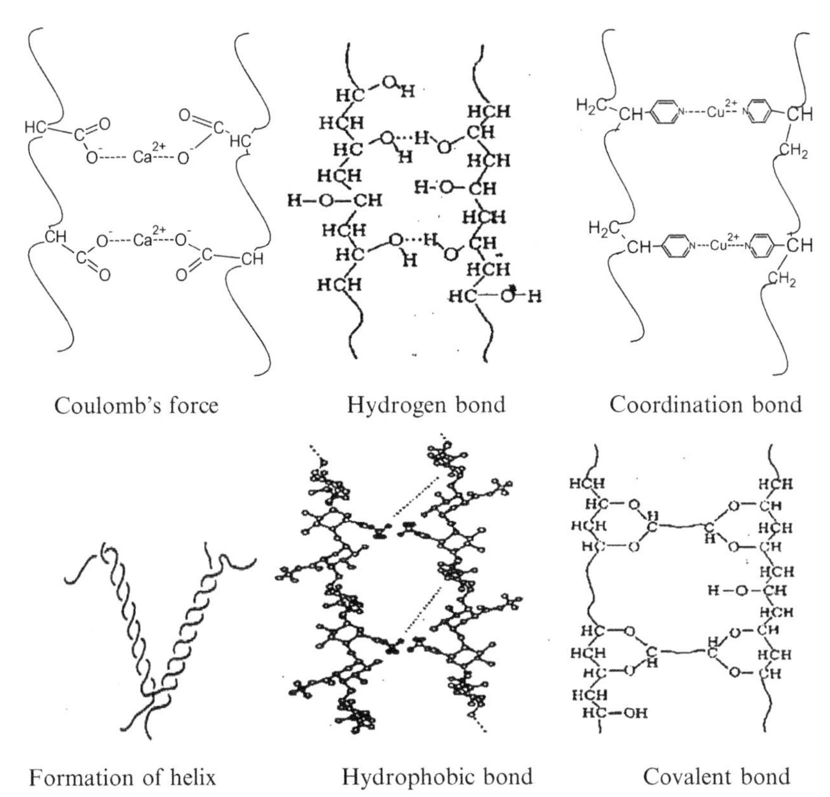

FIG. 1.3 Different types of cross linkage in gel polymer.

the gel system is likely to provide channels for ion migration. The cross-linking occurs either physically or chemically depending upon the conditions and functional groups present on the polymer chains. As the polymer networks are solvated by a large amount of the trapped solvent, so gels generally possess high ionic mobility (Fig. 1.3).

GBPE incorporates both the diffusive property of liquids and the cohesive property of solids. GBPE possesses high ionic conductivity, low volatility, low reactivity, and good operational safety as well as good chemical, mechanical, photochemical, electrochemical, and structural stabilities. It is also lightweight and solvent-free while possessing a wide electrochemical window, high energy density, and good volumetric stability. It is also easily configured into a desired size and shape. GBPE has enhanced safety compared to liquid electrolytes, making it suitable for application in electrochemical devices. Hence, the use of GBPE in batteries prevents leakage and internal shorting and therefore lengthens shelf life. The few drawbacks are poor mechanical strength due to the liquid that is present, which limits its use in wider applications. Nevertheless, this undesirable effect can be eliminated by incorporating the fillers or nanomaterials [29].

1.8.2 Sol-Gel (Gelation)

Biopolymers such as polyesters are interconnected by covalent and coordinate bonding with metals or intermolecular bonding of metals between polymer matrixes. When cooled by the addition of metal ions, a biopolymer forms sol extending for 1 nm–1 μm. Common biopolymers such as starch, chitosan, dextran, alginate, and gelatin when treated with nanoparticles of various metal oxides under controlled atmosphere to get sol-gel. The insolubility of chitosan in basic solutions can be exploited to add another dimension to sol-gel synthesis in the formation of gel beads. In a similar method to the chitosan synthesis of $YBa_2Cu_4O_8$, alginate has been used to prepare nanowires of YBCO superconductors by mixing Y, Ba, and Cu acetate/tartrate salts with alginate to form a gel and then heating to 920°C in air. The sol-gel method of gelation is also applied to prepare biopolymer electrolytes. The presence of nanoparticles improves the stability and redox property of the gel and makes its suitable for energy devices.

1.8.3 Conductivity

The gel polymer electrolytes are prepared by heating a mixture containing the appropriate amounts of the polymer, solvents, and lithium salt to about 120–150°C. This range of temperature is above the glass transition temperature of the polymer in order to form viscous clear liquids. Gel films are made in hot condition by solution casting and allow the solution to cool slowly under the pressure of electrodes. Commonly used plasticizers are less-evaporating solvents for gel polymer electrolytes, such as ethylene carbonate (EC), propylene carbonate (PC), dimethyl formamide (DMF), diethyl phthalate (DEP), di-ethyl carbonate (DEC), methylethyl carbonate (MEC), dimethyl carbonate (DMC), g-butyrolactone (GBL), glycolsulfite (GS), and alkyl phthalates. The solvents have been used separately or as mixtures. Gels are able to retain up to 80% of solvents trapped in the polymer matrix. The use of high permittivity solvents allows a greater dissociation of the lithium salt and increases the mobility of the cation.

In polymer gel electrolytes, the salt generally provides free/mobile ions that take part in the conduction process. The solvent helps in solvating the salt and acts as a conducting medium while the polymer is reported to induce mechanical stability by increasing the viscosity of the electrolyte. Dissociation of Li salt and migration of Li^+ in the gel polymer matrix takes place by means of complex formation between the polar group in a polymer chain and Li^+. A plasticized electrolyte is basically a gel electrolyte, but is unusually associated with the addition of small amounts of a high dielectric constant plasticizer to a salt doped polymer electrolyte to enhance its conductivity. This kind of property gives rise to the possibility of synthesizing good ion-conducting gel materials. The donor number of the polymer repeat units versus that of solvent determines the interaction between solvent-cation or polymer-cation. In Fig. 1.4A, the solvent

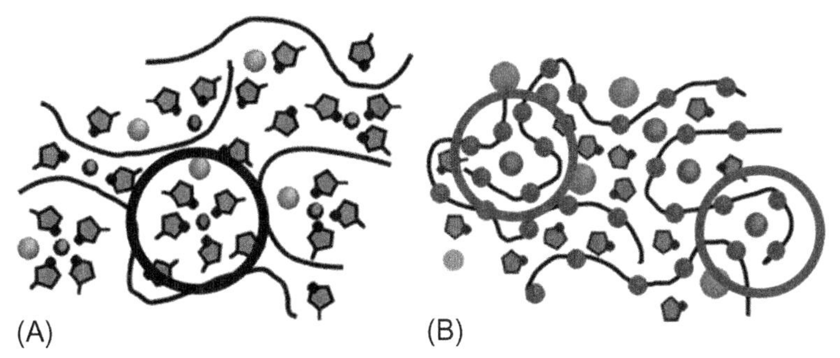

FIG. 1.4 (A) Direct interaction between solvent-cation ($DN_{solvent} > DN_{polymer}$) and (B) direct interaction between polymer-cation ($DN_{polymer} > DN_{solvent}$).

molecule interaction is due to the high DN of the solvent in the gel system, herein, solvent helps the ions move through the polymer chain. In Fig. 1.4B, the higher DN of the polymer causes direct interaction with the ions.

1.9 HYDROGEL BIOPOLYMER ELECTROLYTES (HBPE)

1.9.1 Introduction of HBPE

Plasticizers play an important role in bringing about gelation, along with salts. The variation of plasticizers alters the glass-transition temperature of GBPEs. Hydrogels are a three-dimensional polymeric matrix having water as the plasticizer. Based on the physical structure, hydrogels are further classified as amorphous, semicrystalline, hydrogen-bonded structures, supramolecular structures, and hydrocolloidal. Alternatively, depending on the nature of the polymer, hydrogels are classified as homopolymeric or copolymeric hydrogels.

1.9.2 Mechanism for the Formation of Hydrogel

A hydrogel network is stabilized by physical entanglements, electrostatic attractive forces, and hydrogen bonding. The schematic diagram of a chemical hydrogel with point cross-links and a physical hydrogel with semicrystalline zones is shown in Fig. 1.5A and B, respectively. Hydrogels are viscoelastic solids because they are thermally reversible and depend on the external environment. The factors such as ionic strength, pH, temperature, and electromagnetic radiation influence the swelling of hydrogels.

For example, poly(vinyl alcohol) (PVA), is a cheap, nontoxic, and chemically stable biopolymer. Under acidic conditions, —OH groups of PVA can be reacted with —CHO groups of certain aldehydes to form acetal or hemiacetal linkages. The resultant polymeric entity with acetal or hemiacetal linkages is water-insoluble and gel-like in nature. It can be cast to form thin large surface-area membranes that are used as an electrolyte-cum-separator in

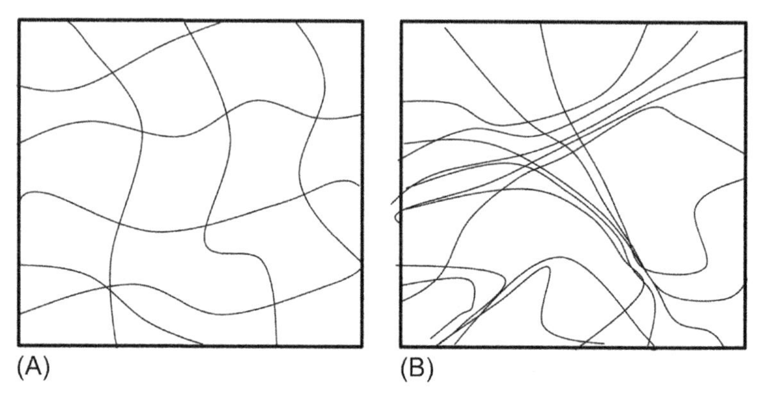

FIG. 1.5 (A) Chemical hydrogel with point cross-links, (B) physical hydrogel with semicrystalline zones.

supercapacitors. Blends of acidic, alkaline, and neutral poly(vinyl alcohol/poly (acrylic acid) blend hydrogel studies show remarkable ionic conductivity. Hydrogels from synthetic polymers have high structural integrity and good mechanical properties. Hydrogels have a large water content absorbed in the polymer matrix that helps in fine-tuning their ionic conductivity. Such hydrogels, however, possess a high degree of hydrophobicity due to the presence of long carbon-chains with fewer hydrophilic pendant groups. By contrast, hydrogels consisting of natural polymers, such as gelatin, possess a high degree of hydrophilicity. This helps in water retention in the polymer matrix, thereby leading to enhanced conductivity. Hydrogel biopolymer electrolytes find their application in biomedical devices such as defibrillators [30]. Gluteraldehyde cross-linked poly(vinyl alcohol)/poly(acrylic acid) blend hydrogels are used as a coating on graphite electrodes for use in electrochemical sensors [31,32].

Poly(ethylene oxide) hydrogel biopolymer electrolytes doped with KOH are used as a biopolymer electrolyte. The disadvantage of this HBPE is that PEO is highly crystalline when dried and has a low melting point, which limits its use in a wide range of temperatures.

Potassium poly(acrylate) hydrogel biopolymer electrolytes with basic formula —$(CH_2\text{-}CHCOOR)_n$—, dissolve in water and the resultant solution will act as hydrogel electrolytes. The biopolymer component of this HBPE is hydrophobic in nature, thereby improving the compatibility between the binder and the electrode matrix.

Gelatin hydrogel biopolymer electrolytes have gelatine as the main component, mostly of proline, hydroxy-proline, and glycine. It is extracted from denatured collagen. When dissolved in boiling water, gelatin forms a pale yellow, semitransparent, viscous solution. Water molecules are bound to polymer chains and hence protonic conduction in pristine gelatine-based HBPE occurs by a Grotthus-type mechanism.

Inorganic hydrogel biopolymer electrolytes are usually sol-gels prepared by the sol-gelation method. They are known for high proton-conducting composites. When compared with organic backbone-based biopolymers, few silica hydrogels incorporating perchloric acid and heteropoly acid dopants have been found to possess ionic conductivity values as high as $10^{-2}\,S\,cm^{-1}$ under ambient conditions. These inorganic hydrogels are not completely biodegradable as they contain metal oxide of various combinations.

1.10 COMPOSITE BIOPOLYMER ELECTROLYTES (CBPE)

CBPEs are prepared to overcome some of the disadvantages of SBPEs. Researchers have manifested many designs of CBPEs such as cross-linking polymer matrices, comb-branched copolymers, doping of nanomaterials, binary salt systems, impregnation with ionic liquids and reinforcement by inorganic fillers, and incorporation of additives (plasticizers). The advantages of CBPEs are good interfacial contact, high thermal stability, good flexibility, and enhanced ion mobility, thereby exhibiting high ionic conductivity. The characteristic of the filler/particles added alters the electronic and ionic conductivities of CBPEs. The characteristics such as particle size, concentration, porosity, surface area, and the interaction between the particles and the polymer matrix are enhanced relative to SBPEs. It also withstands the wear and tear during charging and discharging in electrochemical devices; hence, CBPEs are widely used in energy devices.

In summary, the comparisons of different types of biopolymer electrolytes with their properties are shown in Table 1.2.

1.11 COMPARISON OF SOLID, BLEND, AND GEL BIOPOLYMER ELECTROLYTES

1.11.1 Solid Biopolymer Electrolyte

Advantages

- They are mainly nonvolatile, so provide better durability of the electrochemical devices.
- There is no decomposition at the electrodes.
- There is no possibility of leaks.
- They use metallic lithium in the secondary cells (lithium dendrites growing on the electrode surface would be stopped by the nonporous and solid electrolyte).
- They lower the electrochemical device price. PEO is cheaper than organic carbonates. It could be used as a binder for electrodes to improve the compatibility of consecutive layers. Moreover, fabrication of such a cell would be easier and cost-effective.

TABLE 1.2 Comparison of Biopolymer Electrolytes and Their Properties

Types of Biopolymer Electrolytes	Solid Biopolymer Electrolytes	Blend Biopolymer Electrolytes	Gel Biopolymer Electrolytes	Hydrogel Biopolymer Electrolytes	Composite Biopolymer Electrolytes
Common properties	Lightweight; high-energy density; solvent-free; good flexibility; low volatility; good operational safety; low reactivity and easy to process or configure; good chemical, wide electrochemical window; electrochemical, mechanical, photochemical, volumetric, and structural stabilities				
Specific properties	Easily processed by automation; good mechanical strength	Both properties of polymers will bring good strength of the electrolytes as well as be helpful to the design the devices	High ionic conductivity	Hydrophilic property will be helpful to enhance the ionic conductivity at higher temperatures	Good interfacial contact; high thermal stability; high ionic conductivity
Disadvantages	High interfacial resistance; low ionic conductivity at ambient temperature	Miscibility failure and limited composition of electrolytes	Poor mechanical strength	Difficult to fabricate the device	Poor mixing leading to failure in composites

- Due to the complete solid-state construction, the strength of cells is improved.
- They offer chemical and mechanical stability over a wide temperature range.
- They offer electrochemical stability of at least 3–4 V versus a Li electrode; this is especially important for battery applications.
- The fabrication of electrochemical devices can be made to the desired shape due to the shape flexibility exhibited by polymer electrolytes.
- They lower the cell weight because complete solid-state cells do not need a heavy steel casing.
- They offer improved shock resistance.
- They offer a better overheat and overcharge allowance, therefore improving safety.

Disadvantages

- The solid polymer electrolyte system does not possess any organic liquid, wherein the polymer host is used as a solid solvent, which drastically decreases the conductivity compared to the liquid electrolyte.
- The cycling performance of a dry solid polymer electrolyte with lithium metal electrodes is not satisfactory; as few as 200–300 cycles can be performed.
- The poor performance of the cells is attributed to the poor ionic conductivity of the electrolytes ranging from 10^{-4} to 10^{-8} $S\,cm^{-1}$.

1.11.2 Blend Biopolymer Electrolytes

Advantages

- The engineering plastics are heterogeneous polymer blends with a hard but brittle thermoplastic as the matrix and a soft but tenacious elastomer as the dispersed phase. This morphological principle is the basis for many impact-resistant polymer electrolytes.
- Two hard thermoplastics have gained interest due to their advantages in properties as well as their processability compared to the single components.
- The interfacial bonding between the phases, thus a partial compatibility (partial miscibility) between the matrix and dispersed phase, is of decisive significance for the solid-state properties of heterogeneous blends. For example, the two phases do not separate during processing from the melt.
- Due to the adhesion, the energy-absorbing processes can run across the interphases. Therefore, the shear forces at the surface boundary matrix/dispersed phase arising during an impact (i.e., a rapidly occurring deformation) do not lead to a complete separation of the two phases. Thus a mechanical failure is avoided.

- The thermal stability of the device is improved when a single polymer with poor thermal stability is blended with another having good thermal stability.
- They offer good hydrolytic stabilities and electrical properties and are relatively lightweight compared to their individual polymers.

Disadvantages

- Blending two polymer polymers in a common solvent is time consuming.
- The miscibility of the polymer blends significantly affects the conductivity, hence the proper selection of compatible polymers must be done before blending.
- In some blended polymers such as PVC and PMMA blends, the PVC just provides the mechanical strength to the polymer electrolyte film but does not take part in ion conductivity, hence increasing the resistivity of the polymer electrolyte for ion movement.

1.11.3 Gel Biopolymer Electrolytes

Advantages

- Gel polymer electrolytes attain a high conductivity value approaching that of liquid electrolytes.
- Because of properties such as high ionic conductivity and good adhesion, gel electrolytes have attained precedence over conventional polymer electrolytes.
- They offer low activation energies for conduction.
- They offer high cationic transport numbers.
- They offer good electrode-electrolyte characteristics.
- They offer easy sample preparation.

Disadvantages

- Gel polymer electrolytes have enough mechanical strength, but there still remains the problem of liquid electrolyte due to the phase separation between the polymer matrix, an encapsulated liquid electrolyte.
- They have poor performance at high temperatures, due to entrapped solvents in the gel electrolyte.
- With the continuous charge-discharge process, due to poor intermolecular bonding between the macromolecular chains in the gel polymer electrolytes, the breakage of bonds occurs. Thus, the gelation reduces to a highly viscous liquid, thereby raising the possibility of leakage in electrochemical devices.

REFERENCES

[1] Okada M. Chemical syntheses of biodegradable polymers. Prog Polym Sci 2002;27:87–133.
[2] Chandra R, Rustgi R. Biodegradable polymers. Prog Polym Sci 1998;23:1273–335.
[3] Young AH. Polyvinyl alcohol plasticized amylase compositions. US Patent 1967; 3,312:641.

[4] Tokiwa Y, Suzuki T, Tokiwa Y, Suzuki T. Hydrolysis of polyesters by lipase. Nature 1977;270:76–8.
[5] Gross RA, Kumar KB. Polymer synthesis by in vitro enzyme catalysis. Chem Rev 2001;101:2097–124.
[6] Wright PV. Electrical conductivity in ionic complexes of poly(ethylene oxide). Br Polym J 1975;7:319.
[7] Bruce PG, Vincent CA. Transport in associated polymer electrolytes. New Polym Mater 1990;2:19.
[8] Armand MB, Chabagno JM, Duclot M, Vashisha P, Mundy JN, Shenoy G. Fast ion transport in solids. New York: North Holland; 1979. p. 131.
[9] Mao H, Reamers JN, Jhong Q, von Sacken U. Proceedings of the symposium on rechargeable lithium and lithium ion batteries, 94, 245, the electrochemical society proceeding series, Pennington, NJ; 1995.
[10] Uma T, Mahalingam T, Stimming U. Conductivity studies on poly(methyl methacrylate)-Li_2SO_4 polymer electrolyte systems. Mater Chem Phys 2005;90:245–9.
[11] Stephan AM, Saito Y, Manuel Stephan A, Saito Y. Ionic conductivity and diffusion coefficient studies of PVdF–HFP polymer electrolytes prepared using phase inversion technique. Solid State Ionics 2002;148:475–81.
[12] Conway BE. Electrochemical supercapacitors: scientific fundamentals and technological applications. New York: Kluwer Academic/Plenum Publisher; 1999.
[13] Shmukler LE, Thuc N, Fadeeva YA, Safonova LP. Proton conducting gel electrolytes based on poly(methylmethacrylate) doped with sulfuric acid solutions in N,N-dimethylformamide. J Polym Res 2012;19:9770.
[14] Khandale AP, Bhoga SS, Gedam SK. Study on ammonium acetate salt-added polyvinyl alcohol-based solid proton-conducting polymer electrolytes. Ionics (Kiel) 2013;19:1619–26.
[15] Hamdine M, Heuzey MC, Bégin A. Viscoelastic properties of phosphoric and oxalic acid-based chitosan hydrogels. Rheol Acta 2005;45:659–75.
[16] Tanaka R, Yamamoto H, Kawamura S, Iwase T. Proton conducting behaviour of poly(etheylenimine)-H_3PO_4 system. Electrochim Acta 1995;40:2421.
[17] Gupta PN, Singh KP. Characterization of H_3PO_4 based PVA complex system. Solid State Ionics 1996;86-88:319–23.
[18] Tsuruhara K, Rikukawa M, Sanui K, Ogata N, Nagasaki Y, Kato M. Synthesis of proton conducting polymer based on poly(silamine). Electrochim Acta 2000;45:1391–4.
[19] Arof AK, Shuhaimi NEA, Alias NA, Kufian MZ, Majid SR. Application of chitosan/iota-carrageenan polymer electrolytes in electrical double layer capacitor (EDLC). J Solid State Electrochem 2010;14:2145–52.
[20] Stevens JR, Wieczorek W, Raducha D, Jeffrey KR. Proton conducting gel/H_3PO_4 electrolytes. Solid State Ionics 1997;97:347–58.
[21] He R, Li Q, Xiao G, Bjerrum NJ. Proton conductivity of phosphoric acid doped polybenzimidazole and its composites with inorganic proton conductors. J Membr Sci 2003;226:169–84.
[22] Chandrasekhar V. Polymer solid electrolytes: synthesis and structure. Adv Polym Sci 1998;135:139–205.
[23] Ratner MA, Shriver DF. Ion transport in solvent-free polymers. Chem Rev 1988;88:109–24.
[24] Vincent CA. Ion transport in polymer electrolytes. Electrochim Acta 1995;40:2035–40.
[25] Selvakumar M, Krishna BD. Polyvinyl alcohol–polystyrene sulphonic acid blend electrolyte for supercapacitor application. Physica B 2009;404:1143–7.

[26] Rodríguez J, Navarrete E, Dalchiele EA, Sánchez L, Ramos-Barrado JR, Martín F. Polyvinyl-pyrrolidone–$LiClO_4$ solid polymer electrolyte and its application in transparent thin film supercapacitors. J Power Sources 2013;237:270–6.

[27] Braun D, Cherdron H, Rehahn M, Ritter H, Voit B. Polymer synthesis: theory and practice. 4th ed. New York: Springer; 2005.

[28] Carraher CE. Polymer chemistry. 7th ed. Boca Raton, FL: CRC press, Taylor & Francis Group; 2008.

[29] Ngai KS, Ramesh S, Ramesh K, Juan JC. A review of polymer electrolytes: fundamental, approaches and applications. Ionics 2016;1259–79.

[30] Yao CH, Liu BS, Chang CJ, Hsu SH, Chen YS. Preparation of networks of gelatin and genipin as degradable biomaterials. Mater Chem Phys 2004;83:204–8.

[31] Dasenbrock CO, Ridgway TH, Seliskar CJ, Heinemann WR. Evaluation of the electrochemical characteristics of a poly(vinyl alcohol)/poly(acrylic acid) polymer blend. Electrochim Acta 1998;43:3497.

[32] Shen L, Huang R, Hu N. Myoglobin in polyacrylamide hydrogel films: direct electrochemistry and electrochemical catalysis. Talanta 2002;56:1131–9.

Chapter 2

Methods of Preparation of Biopolymer Electrolytes

Chapter Outline

2.1 SOLID BIOPOLYMER ELECTROLYTE (SBPE)

2.1.1 Polymer Hosts

2.1.1.1 Starch

Starch is a polymer that can be obtained from a wide variety of plants. The major crops used for its production include rice, corn, and potatoes. The starch produced from these plants is in the form of granules. The size and composition

Biopolymer Electrolytes. https://doi.org/10.1016/B978-0-12-813447-4.00002-9

of these starch granules obtained from different plants may vary. Usually, amylose, a linear polymer, constitutes 20 wt.% of the granules and the rest is due to amylopectin, a branched polymer. The average molecular weight of amylose in starch ranges up to 500,000. Although amylose is crystalline, it dissolves in boiling water. Starch is considered a major raw material in film production. This is mainly due to the increasing prices and decreasing availability of conventional film-forming resins. The properties of starch such as its water adsorptive capacity, behavior under agitation and high temperature, and resistance to thermomechanical shear and structural modification have been investigated in detail. When starch is used as an ingredient of biodegradable plastics, it can be melted or blended or just physically mixed with an appropriate polymer. Acetylated starch exhibits many advantages as a structural fiber or film-forming polymer as compared to native starch. Starch acetate is considered to be more hydrophobic has been used for many years as an additive to plastic for various purposes. Starch shows great promise as both a filler and a cross-linking agent in the production of a variety of biopolymers.

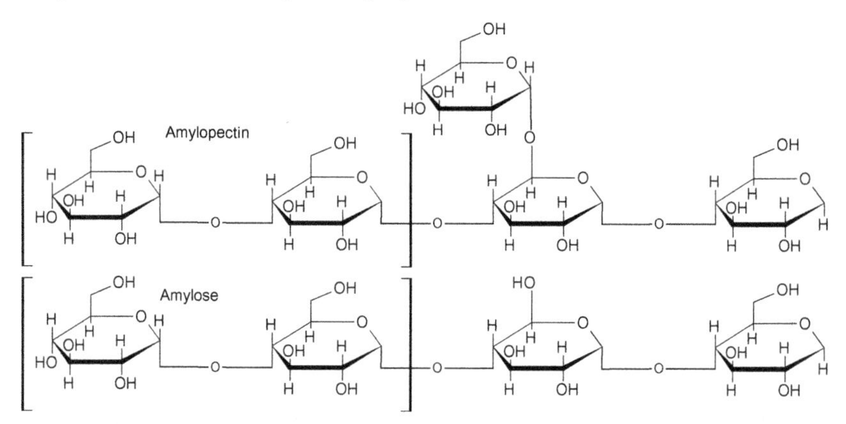

Structure of starch.

2.1.1.2 *Poly(Vinylpyrrolidone) (PVP)*

PVP is a water-soluble polymer. It shows Newtonian viscosity in water. PVP is a light flaky powder under dry conditions and absorbs atmospheric water to an extent of about 18% of its weight. It has very good wetting and film-forming properties in solution form and hence can be used as a good coating or as an additive to coatings.

Although the monomer vinyl pyrrolidone is considered to be toxic to aquatic life, the polymer PVP in its pure form is observed to be very safe. It has been used as a blood plasma expander for trauma victims since the first half of the 20th century. It is harmless and inert to humans and thus is used as a binder in many pharmaceutical tablets. PVP is also used as a base for iodine disinfectant. A PVP-iodine complex is used in the form of solutions, ointment, pessaries, liquid soaps, and surgical scrubs. It is known under the commercial name Betadine. PVP is polar in nature and hence binds to polar molecules exceptionally well. This property of PVP is exploited in its applications as coatings for photo-quality inkjet papers and inks for inkjet printers. PVP is also extensively used in a variety of personal care products such as shampoos, toothpastes, paints, and adhesives for postage stamps and envelopes. It finds use in contact lens solutions and steel-quenching solutions. PVP is used as the base for hair sprays and hair gels. PVP functions as a stabilizer and is used as a food additive. It is also used as a fining agent for white wine in the wine industry. PVP is used as a blocking agent in southern blot analysis as a component of Denhardt's buffer. It is used as an adhesive in glue sticks, battery ceramics, and fiberglass as well as a disintegrant for suspension polymerization. The other applications of PVP are in cathode ray tubes (CRT) as photo resists; in aqueous metal quenching; in the production of membranes for dialysis filters; in agro applications such as crop protection, seed treatment, and coating as a binder and complexing agent; in tooth whitening gels as a thickening agent; and in liquid and semiliquid dosage forms of medicine (syrups, soft gelatine capsules) as an aid for increasing the solubility of drugs.

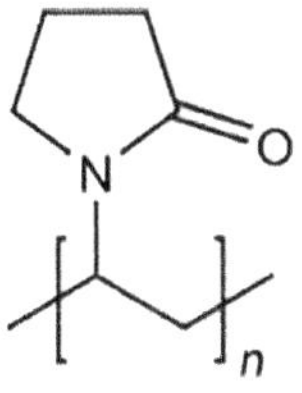

Structure of PVP.

2.1.1.3 Poly(Ethylene Glycol) (PEG)

PEG is used as a dispersant in a variety of toothpastes; it binds water and helps to keep the contents of the toothpaste uniform. It is also under investigation for use as body tattoos for the purpose of diabetes monitoring. PEG is used to create very high osmotic pressures (tens of atmospheres) based on its flexibility and water-soluble property. Further, it has less tendency to have specific interactions with biological chemicals. By virtue of these properties, PEG has become one of the most useful molecules for applying osmotic pressure in biochemistry experiments. PEG is used as the separator and electrolyte solvent in lithium polymer cells. High temperatures of operation may be needed due to its low diffusivity. However, its high viscosity at its melting point facilitates very thin electrolyte layers. PEG is used to deactivate residual phenol in the event of phenol skin burns while working in the laboratory. In gas chromatography, PEG is commonly used as a polar stationary phase. It is also used as a heat transfer fluid in electronic testers. PEG is used as an antifoaming agent in most formulations of soft drinks. PEG is useful in preserving objects that have been recovered from underwater. It makes the wood dimensionally stable and prevents warping or shrinking of the wood by replacing the water in the wooden objects. The PEG precipitation technique for the concentration of viruses is employed in the field of microbiology. PEG is also used as lubricant in certain eye drop formulations. A few ethoxylate derivatives of PEG are used as surfactants. Dimethyl ethers of PEG are the major component of Selexol, a solvent used to remove carbon dioxide and hydrogen sulfide from the gas waste stream in power plants. PEG is coated on drug delivery polymers to make them invisible to white blood cells so as to enable the passage of the medicine throughout the body.

Structure of PEG.

2.1.1.4 Cellulose Acetate (CA)

Wood from trees is a natural raw material and a major source of commercial cellulose. Cellulose constitutes ~50% of the dry weight of wood and is provided in the form of fibers via mechanical and/or chemical pulping methods. Pulping is the purification process used to isolate cellulose and remove most of the natural binders-lignins and hemicelluloses present in the wood structure. Natural cellulose polymer molecules have a molecular weight ranging from 300,000 to 500,000 Da. Each anhydroglucose unit in it has three hydroxyl groups. The chemical modification of this group serves as the basis for most commercially important cellulosic polymers. By far the largest commercial use of fibrous cellulose is in the manufacture of paper and paperboard. Cellulose is easily derivatized, and a wide range of materials can be commercially prepared as a result. There are two major classes of commercial cellulose derivatives: cellulose ethers and cellulose esters. Cellulose ethers find wide application in the food, pharmaceutical, paper, cosmetic, adhesive, detergent, and textile industries.

Cellulose esters can be prepared by either a fibrous or solution acetylation process. The fibrous acetylation process, no longer practiced commercially, involves the strong-acid-catalyzed esterification of the cellulose with an organic anhydride (acetic, propionic, or butyric) in the presence of a nonsolvent diluent. The reaction mixture remains heterogeneous throughout the process. The isolation of the product from the reaction mixture is done through simple filtration. The solution acetylation process is the only commercial process currently practiced for making cellulose esters of organic acids. Solution esterification requires a higher-purity cellulose. Generally, cellulose with a minimal amount of lignin and hemicelluloses impurities and an alpha-cellulose content of at least 95% is required to achieve satisfactory solution processing requirements and product quality standards.

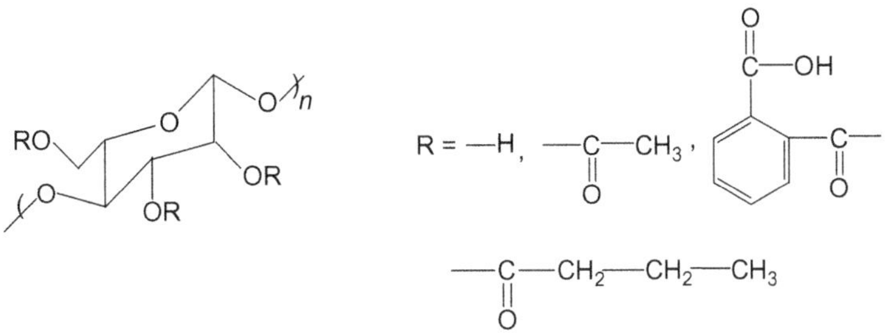

Structure of cellulose acetates.

2.1.1.5 *Poly(Vinyl Alcohol) (PVA)*

PVA is manufactured by the hydrolysis of polyvinyl acetate in the presence of acids or alkalis. PVA, unlike many polymers, is soluble in water. It dissolves in cold water at a slow rate, but the dissolution is fast at higher temperatures, especially around 90°C. PVA serves as an excellent protective colloid for aqueous emulsions and is employed for this purpose in a host of emulsion and suspension systems. It is used in the manufacture of textile fibers where one major aspect to be taken care of is that the fiber as the final product should be made insoluble in water. The fibers made from PVA are found to possess a higher water absorption property in comparison to other synthetic fibers. The fibers are also found to possess excellent resistance to abrasion as well as remarkable tenacity.

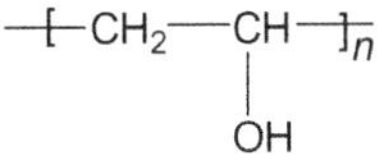

Structure of poly(vinylalcohol).

2.1.1.6 *Chitosan*

Chitosan is a linear polysaccharide composed of β-(1-4)-linked D-glucosamine (deacetylated unit) and *N*-acetyl-D-glucosamine (acetylated unit). It is useful in a number of commercial and biomedical applications. Chitosan is generally derived from the shells of shrimp and other sea crustaceans. Commercially, Chitosan is produced by deacetylation of chitin. Chitin is the structural element in the exoskeleton of crustaceans. The p*K*a value amino group in chitosan is 6.5, thus chitosan is positively charged and soluble in acidic to neutral solutions. The solubility of chitosan also depends on the pH of the solution and the percentage of deacetylation. Chitosan is a bioadhesive and can readily bind to negatively charged surfaces such as membranes. Chitosan and its derivatives such as trimethyl chitosan are used in nonviral gene delivery. In agriculture, chitosan is used as a plant growth enhancer and as a substance that boosts the ability of plants to defend against fungal infections. Chitosan has the ability to rapidly clot the blood. This has made chitosan of use in bandages and other hemostatic agents.

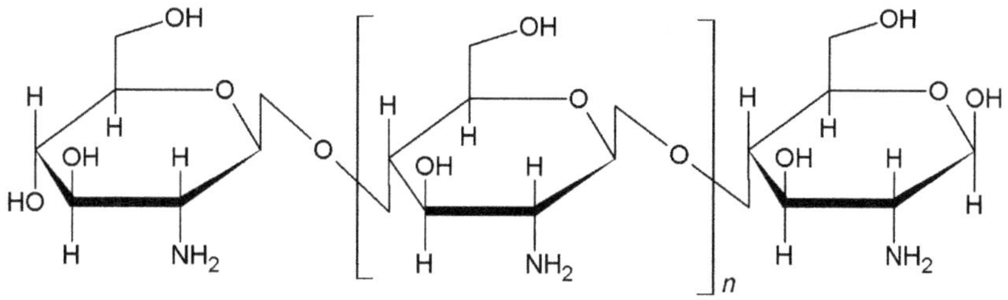

Structure of chitosan.

2.1.1.7 *Poly(Styrenesulphonic Acid) (PSSA)*

PSSA is basically an ionomer based on polystyrene. This polymer dissolves readily in water but is insoluble in lower alcohols. The solid polymer appears as white or off-white in color. PSSA is prepared by the polymerization of sodium styrene sulfonate or by sulfonation of polystyrene. It is used widely in ion-exchange applications and as a super plastifier in cements. This polymer also finds use as a dye-improving agent for cotton and as proton exchange membranes in fuel cell applications.

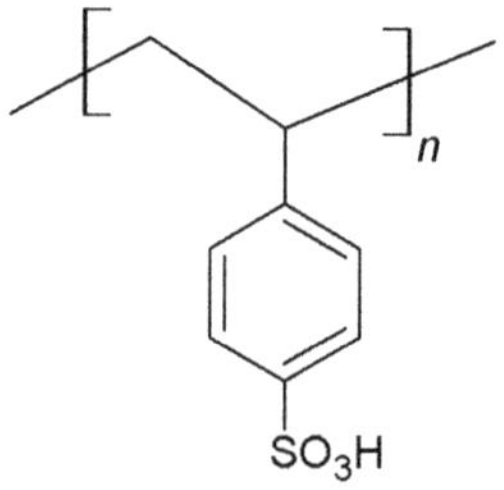

Structure of poly(styrenesulfonic acid).

2.1.2 Transport Properties

It has been established that many salts dissolve in biopolymers of high molecular weight to form ionic conductors and that the solvent plays a major role in the conduction process by facilitating the chain flexibility.

2.1.2.1 Ion-Ion Interactions

It is observed that conductivity of the solution σ increases when a salt is dissolved in a polymer matrix, due to the addition of charge carriers. But when the salt concentration is gradually increased above $\sim$0.1 mol dm^{-3}, the conductivity reaches a maximum and recedes later. This is believed to be because of the formation of transient cross-links in the system, which in turn causes a reduction in the chain mobility. The fall in the conductivity at higher salt concentrations can also be due to the formation of immobile aggregated species in the system [1]. If these species do indeed exist, they are likely to have retarded diffusion rates [2]. It is difficult to distinguish between the effects of ion association and long-range ionic interactions on the reduced ionic mobility.

2.1.3 Solution Casting Method

This is the general procedure for preparing the solid polymer electrolytes. A homogeneous solution of the biopolymer is prepared by dissolving it in aqueous or organic solvents in the presence of the requisite amounts of dopants and plasticizer. The solution is stirred well to ensure uniformity. The solutions are cast on petri plates and dried by maintaining the temperature at $\sim$30°C. The obtained films are used for further studies. Many biopolymer electrolytes are semicrystalline and crystallinity is observed to be affected significantly by the solvent nature, the rate of solvent removal, and by traces of residual solvent that are capable of acting as a plasticizer. In addition, the temperature at which films undergo final drying is important. Sometimes high temperatures induce the formation of high melting of the polymer [3].

2.1.4 Melt Casting Method

To eliminate the effects of solvent, Gray et al. [4] developed a grinding/hot pressing technique for film preparation. This method involves grinding the polymer under liquid nitrogen to a fine powder and subsequently milling requisite amounts of the salt and polymer simultaneously to form a homogeneous mixture.

2.1.5 Plasma Polymerization Method

The plasma polymerization method is suitable for the preparation of ultrathin biopolymer electrolyte films [5]. A layer of plasma-polymerized monomer is first deposited on a stainless-steel, nickel, or gold substrate. This is sprayed with inorganic dopants such as an $LiClO_4$ solution, followed by the deposition of a further polymer layer. The resulting layered structure is heated at 80°C for 24 h under vacuum to produce a homogeneous salt distribution. The plasma parameters are to be optimized to achieve maximum conductivity.

2.2 BLEND BIOPOLYMER ELECTROLYTE (BBPE)

2.2.1 BBPE Host

- A lithium salt-doped plasticized blend biopolymer such as chitosan, starch, PEG [6], or PSSA [7].
- Biopolymer electrolytes consisting of kappa-carrageenan and cellulose derivative blends [8].
- Multiwalled, nanotube-reinforced Chitosan-starch-based electrolytes for batteries.
- Polyvinyl alcohol-polyvinylpyrrolidone poly blend film using a nonaqueous medium [9].
- Fuel cell membrane [10].

2.2.2 Solution Casting Method

This is one of the simplest methods. The solvent must be able to dissolve both the polymers and dopants efficiently. One of the advantage with this method is that drying can be done in simple hot air ovens. The method involves the following steps:

(a) Preparation of polymer and salt solutions in a common solvent.
(b) Mixing of solutions in required ratios.
(c) Uniform stirring of the mixture employing a stirrer or ultrasonic equipment.
(d) Homogeneous casting of the mixture on the required substrate.
(e) Drying in an inert atmosphere under vacuum.

2.2.3 Spin Coating

The spin-coating method is also a simple method and looks similar to solution casting to a certain extent. In this method, instead of direct casting of the film on a substrate, the mixture of polymer and inorganic salt solution is drop cast on a substrate. The substrate is then placed in a spin coater and subjected to rotation at a desired rate. The film thickness is controlled by proper adjustment of the solution as well as the rotation speed. The disadvantage of this method is that highly viscous solutions cannot be used here. Similarly, it is not applicable for gel mixtures, as the mixture droplet cannot be spread into a thin film by the rotation process.

2.2.4 Hot Press

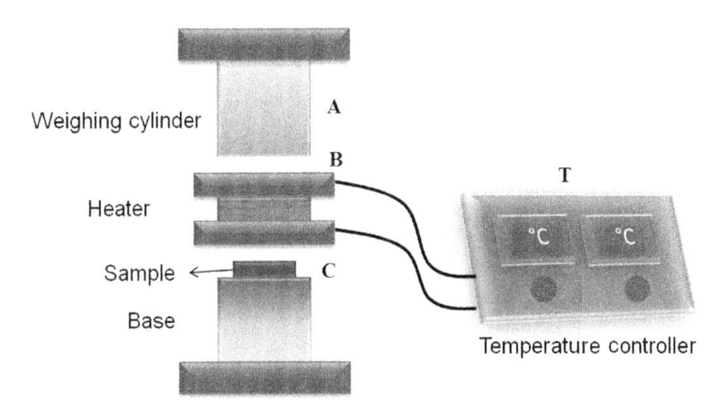

The schematic of the equipment used for the hot press technique is shown in the figure. The components of the equipment are: (**A**) weighing cylinder, (**B**) heating chamber, (**C**) basement, and (**T**) temperature controller. Appropriate amounts of polymers, salt, and filler are mixed for several minutes in a mortar. The polymer powder mixture is then sandwiched between two sheets present inside the heating chamber, which is maintained at temperatures above the melting point of the components of the mixture. The sample is then pressed (5–10 min) with the required pressure, which can be controlled by the weighing cylinder. The sample is slowly cooled to room temperature after enough heating and pressing. The sample is then removed from the heating chamber and stored in a glove box.

2.2.5 Aspects of Conductivity in BBPE

In general, the conductivity of the polymer electrolyte system increases with an increase in the concentration of dopant salt due to the enhancement in the number of charge carriers. The studies on polymer interactions and the influence of temperature on conductivity showed that a rise in temperature causes an increase in the conductivity of the system.

In batteries, the use of a liquid electrolyte can provide high ionic conductivity. But the presence of liquid electrolyte can be disadvantageous to its mechanical strength. This represents a big challenge for good design of batteries and fuel cells employing new multifunctional materials for engineering applications. Many polyelectrolyte systems consisting of lithium salt and synthetic polymers such as poly(ethylene oxide), poly(propylene oxide), poly(bis-methoxy ethoxy ethoxide)-phosphazene, poly(dimethyl siloxane), poly(acrylonitrile), poly(methyl methacrylate), poly(vinyl chloride), and poly(vinylidene fluoride) are being used as polyelectrolytes. Although they possess the required conductivity, they are not easily degradable in the environment. The blending of biopolymers with lithium salt and multiwalled carbon nanotubes (MWNTs) is used to improve both the conductivity and the mechanical strength of the system.

2.3 GEL BIOPOLYMER ELECTROLYTE (GBPE) AND HYDROGEL BIOPOLYMER ELECTROLYTE (HBPE)

2.3.1 Casting Method

In this method, a suitable polymer that can serve as the matrix is dissolved in an appropriate low boiling point solvent. The chosen dopant, which may be an inorganic salt or an acid, is also dissolved in the same solution and stirred well for uniform mixing. The resulting viscous solution is used for casting. The films of gel electrolyte are formed after partial removal of the solvent from the mixture. Generally these gel polymer electrolytes exhibit very good conductivity but are weak in mechanical properties due to the presence of traces of liquid content. Cross-linking is employed to strengthen the polymer matrix.

2.3.2 Phase Inversion Method

The phase separation/inversion method is a very convenient method for preparing microporous membranes. The two processes of this method are (a) precipitation in a nonsolvent, and (b) solvent evaporation evaporation induced.

(a) Precipitation in a nonsolvent

In this process, the required amount of the polymer solution is coated on a glass slide and immersed in a nonsolvent. The polymer precipitates from the solution while the solvent is being replaced by the nonsolvent, and thin porous gel polymer films are produced.

(b) Solvent evaporation-induced precipitation

In this process, the copolymer or polymer is first dissolved in a mixture of volatile acetone as the solvent and ethanol as the nonsolvent. The polymer solution is then cast onto a sheet to form a film. Air drying results in a film with pores. Pentane is used as a nonsolvent to increase the pore size in the formed gel polymer electrolyte. Due to pores, the ionic conductivity in the gel can be increased by this process.

2.3.3 Electrospinning

In this method, electric force is used to draw charged threads of polymer solutions or polymer melts with the formed fiber diameters in the order of nano to micrometers. The formed electrospun gel polymer electrolyte is a suitable material for use as a host matrix in microporous polymer electrolytes because the fully interconnected pores with large surface areas can function as efficient channels for ion conduction.

2.3.4 Sol-Gel Process

The sol-gel process for polymers is summarized in the following key steps and the gelation nature of some biopolymers is also discussed.

Type of Gel	Bonding	Source	Gel Schematic
Polymer complex I In situ polymerizable complex (Pechini method) [11]	Organic polymers interconnected by covalent and coordinate bonding	Polyesterification between polyhydroxy alcohol (e.g., ethylene glycol) and carboxylic acid with metal complex (metal-citrate)	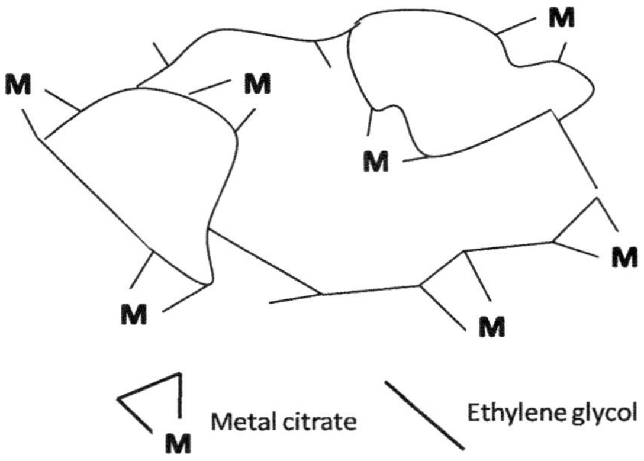 M M M M M M M M M Metal citrate Ethylene glycol

Continued

Type of Gel	Bonding	Source	Gel Schematic
Polymer complex II Coordinate and crosslinking polymers [12]	Organic polymers interconnected by coordinate and intermolecular bonding	Coordinate polymer (e.g., alginate) and metal salt solution (typically aqueous)	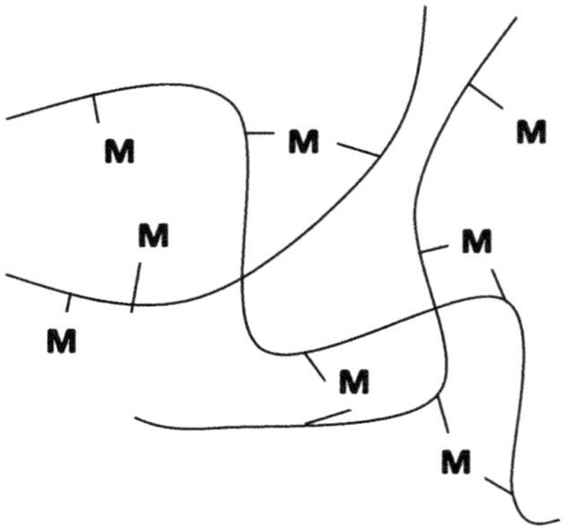 M M M M M M M M M

In sol-gel chemistry, the major requirements for polymers are that they should be soluble in at least one solvent, preferably water, and possess functional groups that can bind to metal ions. Naturally available and water-soluble polymers are ideal resources for this process.

2.3.5 Hydrogel Biopolymer Electrolyte (HBPE)

Hydrogels are polymer networks with water as their solvent; this accumulates between the intermolecular spacing of the polymer. Although hydrogels are generally prepared based on hydrophilic monomers, hydrophobic monomers are also sometimes used in hydrogel preparation to regulate the properties for specific applications. The technique in the polymerization of hydrogels decides the appearance in the form of film, microsphere, or gel. The classification of hydrogels depends on their physical structure and chemical composition; they can be classified as follows:

(a) Amorphous
(b) Semicrystalline (mixture of amorphous and crystalline phases)
(c) Crystalline

According to the electrical charge concept, hydrogels are categorized on the basis of the presence or absence of an electrical charge located on the cross-linked chains, as given below:

(a) Nonionic/neutral
(b) Ionic (anionic or cationic)
(c) Amphoteric electrolyte containing both acidic and basic groups.
(d) Zwitterionic (polybetaines) containing both anionic and cationic groups in each structural repeating unit.

Some of the natural polymers that form hydrogels are collagen, gelatin, starch, alginate, and agarose. Synthetic polymers that form hydrogels are traditionally prepared using chemical polymerization methods. In general, hydrogels can be prepared from either synthetic polymers or natural polymers. The synthetic polymers are chemically stronger compared to the natural polymers. Thus, synthetic hydrogels provide higher durability but a poor degradation rate. Hydrophilic monomers with multifunctional cross-linkers are used in copolymerization, cross-linking, and free-radical polymerization of hydrogels. Water-soluble linear polymers of both natural and synthetic origin are cross-linked to form hydrogels in several ways:

(a) Linking polymer chains via a chemical reaction.
(b) Using ionizing radiation to generate main-chain free radicals that can recombine as cross-link junctions.
(c) Physical interactions such as entanglements, electrostatics, and crystallite formation.

In general, the three main reagents of the hydrogel preparation are monomer, initiator, and cross-linker. Aqueous solutions are used to control the heat during polymerization and the final hydrogel properties, followed by a water wash to remove impurities left from the preparation process. Polar monomers are usually used as the starting material to prepare hydrogels, either natural or synthetic or a combination of both.

2.3.6 Bulk Polymerization

Vinyl monomers are effectively used for the production of hydrogels. Using one or more types of monomers, a wide variety of hydrogels can be prepared using the bulk polymerization technique for a given application. In hydrogel formulation, a small amount of cross-linking agent is added and the polymerization reaction is initiated with radiation, ultraviolet, or chemical catalysts. The polymerized hydrogel may be produced in a wide variety of forms including films and membranes, rods, particles, and emulsions. This technique produces homogeneous hydrogel with a hard glassy transparent polymer matrix that becomes soft and flexible when immersed in water.

2.3.7 Solution Polymerization/Cross-Linking

In this reaction process, the ionic or neutral monomers are mixed with the multifunctional crosslinking agent. Either UV irradiation or a redox initiator system is used for the initiation of polymerization. The prepared hydrogels are washed with distilled water to remove the unreacted monomers, oligomers, the cross-linking agent, the initiator, the soluble and extractable polymer, and other impurities. If the amount of water during polymerization is more than the water content corresponding to the equilibrium swelling, there will be phase separation and the heterogeneous hydrogel is formed. The common solvents used for solution polymerization of hydrogels are water, ethanol, benzyl alcohol, and water-ethanol mixtures. The gel is made to swell in water to remove the solvent.

2.3.8 Suspension Polymerization

In this technique, the reagents such as monomers and initiators are dispersed with continuous agitation and the addition of a low hydrophilic-lipophilic-balance (HLB) suspending agent. In most cases, the water-in-oil (W/O) process is chosen instead of the more common oil-in-water (O/W). With the dispersion being thermodynamically unstable, the polymerization occurs quickly and suspends as microspheres or powder. This process is advantageous since no grinding is required for the prepared polymers.

2.3.9 Grafting to a Support

In this process, polymers that are weak in structure are grafted with strong polymers. The free radicals generated are made to polymerize on stronger ones, thus increasing the chain length. Usually, this technique is used for improving the strength of the polymer obtained from the bulk polymerization.

2.3.10 Polymerization by Irradiation

In this process, the initiator used is high energy radiation such as gamma rays and electron beams. Pure and chemical initiator-free hydrogels are obtained in this process. When the aqueous polymer solution is irradiated with high-energy radiation, free radicals on polymer chains are generated and hydroxyl radicals are generated from water molecules. These hydroxyl radicals also attack the polymer chain to produce macroradicals, which combine covalently to get a cross-linked hydrogel structure.

2.3.11 Technical Features of Hydrogel

Hydrogel as an ideal biopolymer has the following features:

- The highest absorbency under load.
- The desired rate of absorption (preferred particle size and porosity) depending on the application requirement.
- The highest biodegradability without the formation of toxic species following degradation.
- The highest absorption capacity (maximum equilibrium swelling) in saline.
- The lowest soluble content and residual monomer.
- The highest durability and stability in the swelling environment and during storage.
- The lowest price.
- Photo stability.
- pH-neutrality after swelling in water.
- Colorless, odorless, and absolutely nontoxic.
- Rewetting capability: the hydrogel has to be able to give back the imbibed solution or to maintain it; depending on the application requirement

Natural polymer-based hydrogels could function as a porous structure with a large water content, providing a new "structure" for certain reactive and polymerizable monomers such as metal nanoparticles and ions, inorganic nanoparticles, graphene, carbon nanotubes (CNTs), graphene oxide, conducting polymers (CPs), and ionic liquids. Under certain conditions, the monomers would interact and then bond with the polymer chains of the hydrogels, or improve such existing properties. Certain hydrogels could be prepared using a sol-gel transition process where electroactive materials are mixed with a natural polymer solution. Therefore, it could be possible for electroactive materials to be directly dispersed into hydrogels.

2.4 COMPOSITE BIOPOLYMER ELECTROLYTES (CBPE)

The composite biopolymer electrolytes have overcome the disadvantages and limitations of other polymer electrolytes because they have good flexibility, good interfacial contact, high thermal stability, and high ionic conductivity by enhancing ion mobility. The inorganic salts added to enhance the electronic and ionic conductivities of CBPEs depend on porosity, particle size, surface area concentration, and their interaction with the polymer matrix.

Preparation of CBPE involves techniques such as cross-linking polymer matrices, polymer blending, branched copolymers along with incorporation of additives such as plasticizers, reinforcement by inorganic fillers, binary salt systems, and impregnation with ionic liquids doping of nanomaterials.

2.4.1 Polymer Blending Method

In this process, two polymers are mixed in a common solvent and a solution containing inorganic salts is added in the required amounts. They are mixed to form a homogeneous slurry and a given amount of dopants is added with stirring. The prepared solution is casted and vacuum-dried under a specific temperature for a few days.

2.4.2 Cross-Linking of Polymer Matrices

Composite biopolymer electrolytes can be prepared by using cross-linking agents by the solvent casting technique. Normally glutaraldehyde, borax, and 2,2′-azobis(isobutyronitrile) are used as a cross-linking reagent to prepare any type of polymer electrolyte.

2.4.3 Incorporation of Additives and Plasticizers

Plasticizers are additives that enhance plasticity and fluidity as well as aid in dissolving inorganic salts in the polymer matrix. This plasticizer provides a channel for ionic conductivity in the films. Hence, biopolymer electrolytes with plasticizers have a relatively higher conductivity than solid biopolymer electrolytes. Common plasticizers such as propylene carbonate, diethylene carbonate, and ethylene carbonate play an important role in providing sufficient mobility or ionic conduction. The main drawbacks of using plasticizers are that they give poor mechanical properties and some polar plasticizers react with the inorganic ions.

2.4.4 Doping With Nanomaterials

Doping with nanomaterials proved to be effective in improving mechanical properties along with enhanced ionic conductivity; even the interface stability with various electrode materials has improved. Usually, nanomaterials are ceramic fillers such as TiO_2, γ-$LiAlO_2$, SiO_2, and $BaTiO_3$; they are used to obtain composite polymer electrolytes.

2.4.5 Impregnation With Ionic Liquids

Natural polymers such as polysaccharides have good solubility for ionic liquids as they form hydrogen bonding with the polymer. Solubility can be affected by both the cation and anion of the ionic liquid. The unique properties of ionic liquids make them an ideal solvent for the preparation of composite biopolymer electrolytes.

2.4.6 Reinforcement by Inorganic Fillers

Normally, inorganic fillers and additives are bound together with a polymer matrix to form the composite biopolymer electrolyte. A large number of fillers in a variety of forms such as spheroidal, fibrous, or porous, along with other additives, have been studied to enhance the performance of the composites. A poly(acrylic acid) filler-based biopolymer electrolyte was prepared by Chiam-Wen et al. [13]. A biopolymer doped with a lithium salt solution is initially dissolved in distilled water and then the required amount of filler is added to the solution. The resulting solution is subjected to sonication for a few hours at 60–70°C and stirred overnight at 50°C. The solution is then cast on a petri dish and dried in an vacuum oven at 50°C to form the composite biopolymer electrolyte.

REFERENCES

[1] Archer WI, Armstrong RD. Stability of some lithium ion conducting polymeric solid electrolytes in the presence of lithium electrodes. Electrochim Acta 1981;26:167.

[2] Fauteux D, Prud'Homme J, Harvey PE. Electrochemical stability and ionic conductivity of some polymer-lix based electrolytes. Solid State Ionics 1988;28–30(Part 2):923–8.

[3] Martin SW. Conductivity relaxation in glass: compositional contributions to non-exponentiality. Appl Phys A Solid Surf 1989;49(3):239–47.

[4] Gray FM, MacCallum JR, Vincent CA. Poly(ethylene oxide)—$LiCF_3SO_3$—polystyrene electrolyte systems. Solid State Ionics 1986;18–19(Part 1):282–6.

[5] Yasuda H, Gazicki M. Biomedical applications of plasma polymerization and plasma treatment of polymer surfaces. Biomaterials 1982;3(2):68–77.

[6] Sudhakar YN, Selvakumar M, Bhat DK. $LiClO_4$-doped plasticized chitosan and poly(ethylene glycol) blend as biodegradable polymer electrolyte for supercapacitors. Ionics (Kiel) 2012; 19(2):277–85.

[7] Sudhakar YN, Selvakumar M. Ionic conductivity studies and dielectric studies of poly(styrene sulphonic acid)/starch blend polymer electrolyte containing $LiClO_4$. J Appl Electrochem 2012;43(1):21–9.
[8] Rudhziah S, Rani MS, Ahmad A, Mohamed NS, Kaddami H. Potential of blend of kappa-carrageenan and cellulose derivatives for green polymer electrolyte application. Ind Crop Prod 2015;72:133–41.
[9] Rajeswari N, Selvasekarapandian S, Karthikeyan S, Prabu M, Hirankumar G. Conductivity and dielectric properties of polyvinyl alcohol—polyvinylpyrrolidone poly blend film using non-aqueous medium. J Non-Cryst Solids 2011;357(22 – 23):3751–6.
[10] Acar O, Sen U, Bozkur A, Ata A. Blend membranes from poly (2, 5-benzimidazole) and poly (styrene sulfonic acid) as proton-conducting polymer electrolytes for fuel cells. J Mater Sci 2010;45:993–8.
[11] Lin J, Yu M, Lin C, Liu X. Multiform oxide optical materials via the versatile Pechini-type sol – gel process: synthesis and characteristics. J Phys Chem C 2007;111(16):5835–45.
[12] Hall SR. Biomimetic synthesis of high-Tc, type-II superconductor nanowires. Adv Mater 2006;18:487–90.
[13] Liew C-W, Ng HM, Numan A, Ramesh S. Poly(acrylic acid)–based hybrid inorganic–organic electrolytes membrane for electrical double layer capacitors application. Polymer 2016; 8(5):179.

Chapter 3

Biopolymer Electrolyte for Supercapacitor

Chapter Outline

3.1 INTRODUCTION OF ELECTROCHEMICAL CAPACITOR (SUPERCAPACITOR)

An electric/electrochemical double-layer capacitor (EDLC) or supercapacitor or ultracapacitor is a unique electrical storage device that can store much more energy than conventional capacitors while offering much higher power density than batteries. EDLCs fill the gap between batteries and the conventional capacitor, allowing applications for various power and energy requirements, that is, backup power sources for electronic devices, load leveling, engine start or acceleration for hybrid vehicles, and electricity storage generated from solar or wind energy.

Biopolymer Electrolytes. https://doi.org/10.1016/B978-0-12-813447-4.00003-0

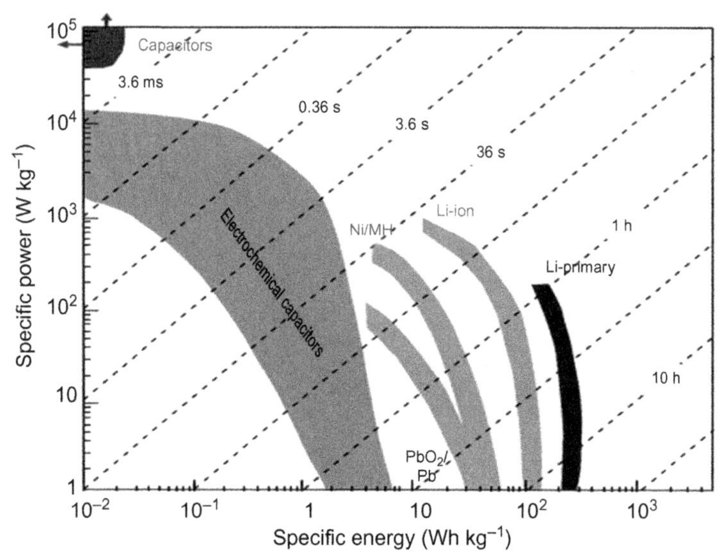

FIG. 3.1 A Ragone plot showing the performance of different energy devices [1]. *(Reproduced from Nature publishing group.)*

Fundamentally, electrical energy is stored in two different ways in energy devices: (a) indirectly, with chemical energy converting to electric energy generated from electroactive materials undergoing a faradic redox reaction between two electrodes having different electrode potentials, and (b) directly, with non-faradic, that is, electric energy stored in an electrostatic way between negative and positive electrodes separated by a dielectric medium.

Electrochemical cells are devices that convert chemical energy into electrical energy and vice versa. Due to the redox reaction in the electrode-electrolyte interface, this electrical energy can be obtained. The electrochemical cells mainly consist of a negative electrode and a positive electrode having active mass and an ion-conducting electrolyte.

In response to the changing global landscape, energy has become a primary focus of the major world powers as well as the scientific community. There has been great interest in developing and refining more efficient energy storage devices. One such device, the capacitor, has matured significantly over the last three decades, emerging with the potential to facilitate major advances in energy storage. Fig. 3.1 represents the energy density and power density distribution curves of the electrochemical capacitor with comparison to other energy devices. This plot is called a Ragone plot.

3.1.1 Taxonomy of Supercapacitors

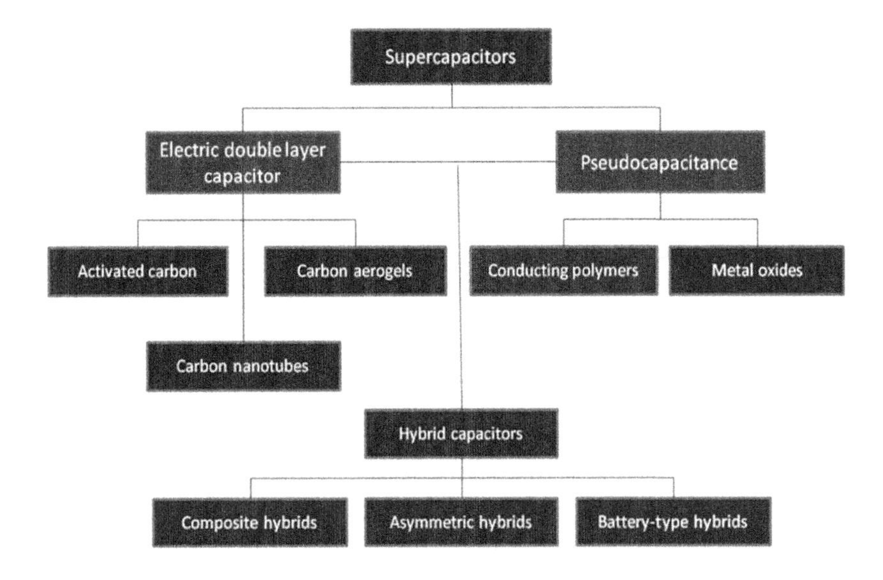

3.1.2 Brief Evolution of Capacitor to Supercapacitor

A capacitor stores energy at the electrode/electrolyte interface due to the formation of an electrochemical double layer; it is currently called by several names under the trade name and the capability of capacitors. Names such as supercapacitor, ultracapacitor, or electrochemical double-layer capacitor are often used. Also, because there is an additional contribution to the capacitance other than the double-layer affects, it is called an electrochemical capacitor. In this book, the technology is referred to as "supercapacitor" but based on the mechanism of charge storage, sometimes the term "electrochemical double-layer capacitor" (EDLC) is used.

3.1.2.1 First-Generation Capacitors From Condensers

Early capacitors were also known as condensers, a term that is still occasionally used today. The term was first used for this purpose by Alessandro Volta in 1782, with reference to the device's ability to store a higher density of electric charge than a normal isolated conductor.

In a capacitor, energy is stored via an electrostatic field generated from the removal of charge carriers, typically electrons, from one metal plate, then depositing them on another. Hence, it is called an electrostatic capacitor. Potential is developed between two plates because of this charge separation and therefore energy can be harnessed from the capacitor. This energy stored will be proportional to both the number of charges accumulated and the potential difference between the plates. Hence, charge storage depends on the size and

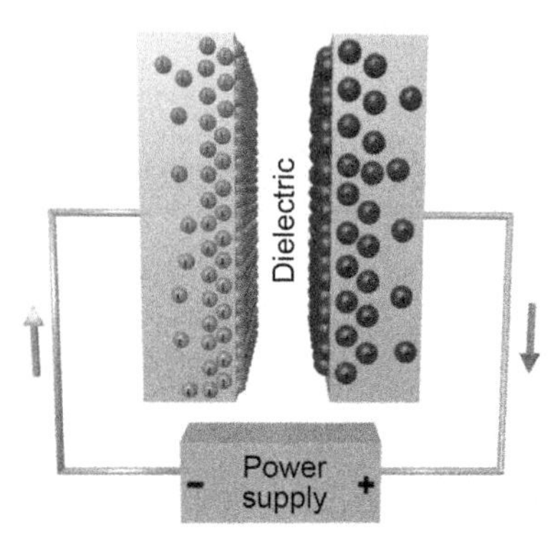

FIG. 3.2 Schematic representation of energy stored in an electrostatic capacitor.

material properties of the plates, and voltage developed depends on the dielectric breakdown. The power densities of a capacitor are higher than other energy storage devices, but it still has poor energy density. Nevertheless, optimizing the dielectric materials suitable to the size of capacitor can lead to higher energy densities. Electrostatic capacitors, shown schematically in Fig. 3.2, consist of a pair of conductors separated with a dielectric such as air, mica, polymer film, ceramic, etc. These capacitors operate at gigahertz (GHz) frequencies with charge times of $\sim 10^{-9}$ s.

3.1.2.2 Second-Generation Electrolytic Capacitors

Electrolytic capacitors are made up of tantalum, ceramic, and aluminum, either containing a solid or liquid electrolyte separated by a separator. The construction of the cell is similar to that of batteries.

3.1.2.3 Third-Generation Electrochemical Double-Layer Capacitors

The capacitor evolved as an electric/electrochemical double-layer capacitor (EDLC) in the third generation. The main focus was on the electrical charge at the electrode-electrolyte interface in order to reach $\sim 10^6$ F. In this generation, organic and aqueous electrolytes were used in the construction of the cell and activated carbon was primarily exploited as the electrode material. The power density and low energy density output from these EDLCs were complementary to batteries. Being a modified capacitor, EDLC has a longer life cycle and power density than batteries. They even possess a higher energy density as compared to conventional capacitors. Thus, due to this hybrid concept between battery and conventional capacitor, EDLC has been used in the engine start or acceleration for hybrid vehicles as well as backup power sources for electronic devices, load leveling, and electricity storage generated from solar or wind energy. The practical use of EDLCs began in 1957 by General Electric, which sought a patent for an electrolytic capacitor consisting of four porous carbon electrodes [2]. Later, many companies successfully commercialized supercapacitors having a capacity from 7 to 4000F and voltage from 2 to 6V.

Various flexible supercapacitors have also been prepared with suitable energy density. Flexible supercapacitors have great potential for applications in wearable, miniaturized, portable, large-scale transparent and flexible consumer electronics. Flexible electrodes are primarily based on carbon material because carbon networks such as carbon fabric, carbon film, carbon paper, and carbon textile can be used for electrochemical performances. The electrolyte component of a flexible supercapacitor is either liquid electrolyte or polymer electrolyte. Liquid electrolyte varies from aqueous to organic solutions. A mixture of organic solution and salts is widely used. The polymer electrolyte is generally a mixture of gelling agent, solute, and solvent; it has great potential in future transparent flexible energy storage devices due to its safety and no need of a separator [3,4].

3.2 PRINCIPLE

The principle of EDLC works on the development of double-layer capacitance at the electrode/electrolyte interface. Specific capacitance can be defined by various ways from double-layer capacitors, such as normalization to unit surface area, volume, or mass of the device. However, because the focus of this perspective is materials, capacitance is normalized to unit mass of the active material, which simplifies comparisons between materials. It is important to emphasize that because supercapacitor voltage decreases linearly with the state of charge, not all the stored energy can generally be used. The energy density and power density depend upon the equivalent series resistance (ESR). ESR is generated due to the electrolyte resistance, electrode resistance, and resistance due to the diffusion of ions in the electrode porosity.

3.2.1 Formation of the Electrical Double Layer

A simple EDLC is shown in Fig. 3.3 in which two carbon rods are dipped in the salt water. The dissociated ions are dispersed well in the solution when there is no current applied to the cell. When the current is applied between the rods, current starts flowing and charge accumulates on the electrode/electrolyte interface. An excess or a deficit of electric charge is accumulated on the electrode surfaces and an equal counterbalancing of opposite charge accumulates on the electrolyte end. When the switch is opened, the voltage remains due to the stored energy. The energy is from the double-layer capacitor developed in series. These two parallel regions of charge form the source of the term "inner Helmholtz double layer," The thickness of the double layer depends on the concentration of the electrolyte and the size of the ion, which varies from

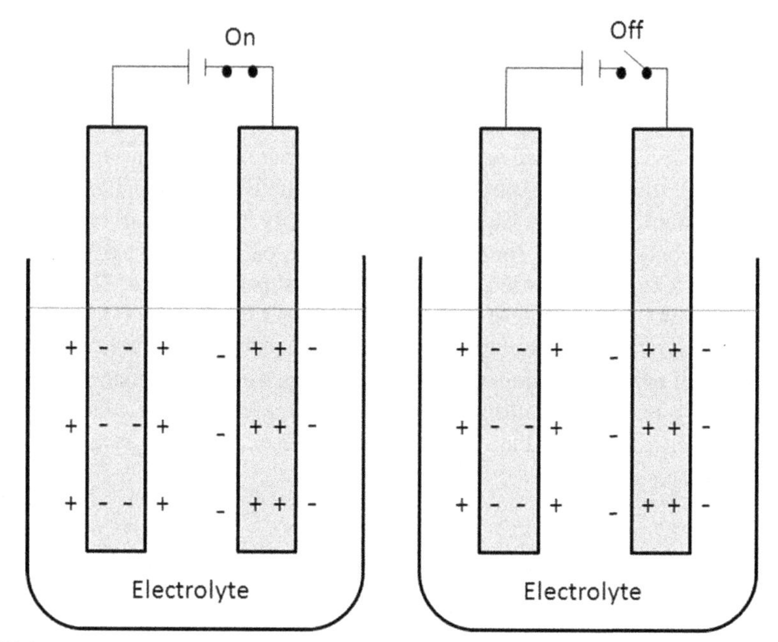

FIG. 3.3 Electrochemical double-layer capacitor showing the persistence of the double layer at the interface after the switch is opened.

a few angstroms. Hence, the surface area is measured in thousands of square meters per gram ($m^2\ g^{-1}$) of electrode material.

As described by Helmholtz, from this inner double layer the double-layer capacitance C is calculated using the equation:

$$C = \varepsilon_0 \varepsilon_r \frac{S}{D} \tag{3.1}$$

where ε_0 is the electric constant (8.854×10^{-12} F m^{-1}), S is the specific surface area of the electrodes ($m^2\ g^{-1}$), ε_R the relative dielectric constant of the interface (whether liquid or solid), and D(m) is the separation of the electrode plates.

3.2.2 Stern Theory of the Double Layer

Furthermore, this capacitance model was later refined by Gouy and Chapman and Stern and Geary, who suggested the presence of a diffuse layer in the electrolyte due to the accumulation of ions close to the electrode surface, as described in Fig. 3.4. The Stern model shows the inner Helmholtz plane (IHP) and the outer Helmholtz plane (OHP). The IHP refers to specifically

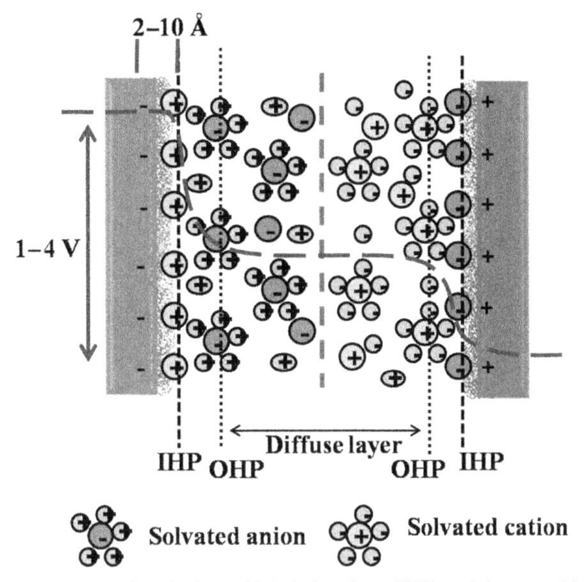

FIG. 3.4 Stern model showing the inner Helmholtz plane (IHP) and the outer Helmholtz plane (OHP), the potential drop at the electrode/electrolyte interface and diffuse layer.

adsorbed ions and the OHP refers to nonspecifically adsorbed ions. The diffuse layer begins from the OHP layers toward the bulk of the EDLC.

Using a series of EDLC high capacitances with the desired voltage can be achieved. This is due to the use of high surface area porous electrodes, which makes the EDLCs slower than conventional capacitors. Fig. 3.5 shows electrical double-layer capacitance coming from the interface between the electrode material particles (such as activated carbon) and the electrolyte.

The effectiveness of the increased capacitance due to nanoporous carbon is represented in Fig. 3.6. In this high surface area represented in the cylinder, the electrolyte fills and forms electric double layer on the interior wall surface of the pore, which makes capacitor to connect in series with each other and hence increasing the capacitance of the cell [5].

The conductivity in carbon is greater than the electrolyte, so the electrical signal passes through the inner side of the cylinder faster than through the pores in the electrolyte. The charge stored on the pore mouth is accessible with small electrolyte resistance while the charge stored within the pore takes a longer time and path with high series resistance. This limits the EDLC up to 1–3 V compared to a conventional capacitor.

The main feature of the EDLC is that no charge transfer takes place across the electrode and electrolyte interface. This process is called the nonfaradaic process. This implies that the concentration of the electrolyte remains constant during the charge/discharge cycle.

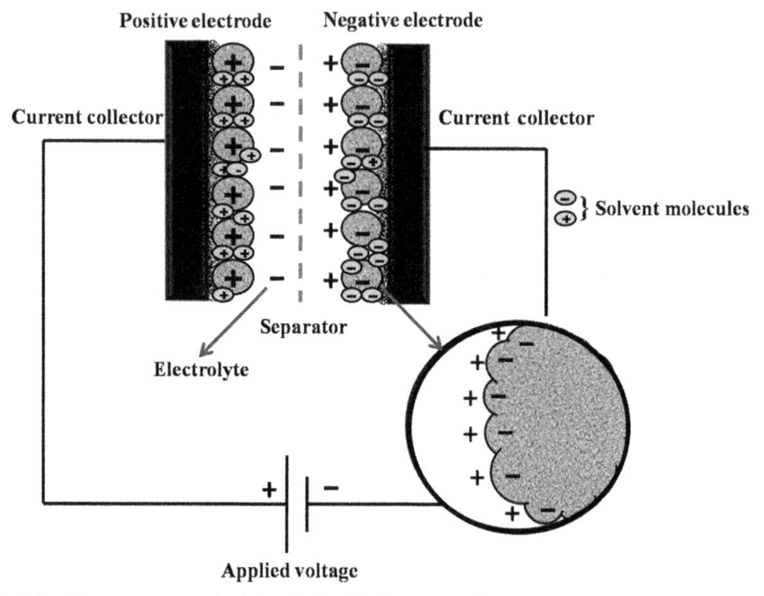

FIG. 3.5 Charge storage principle of a double-layer capacitor.

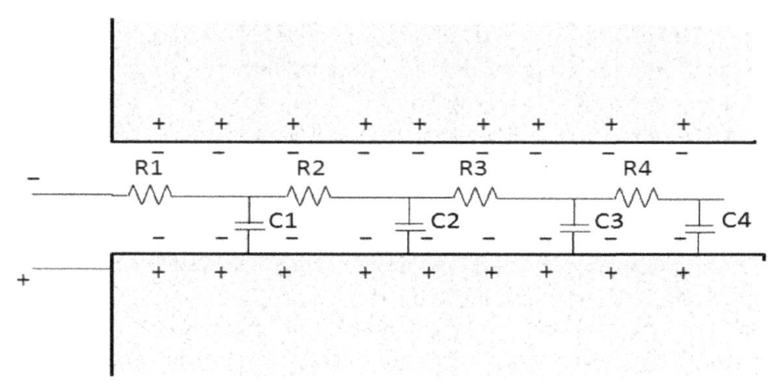

FIG. 3.6 Cylindrical representation of a nanopore in a carbon electrode of an electrochemical capacitor.

3.2.3 Mechanism

Let the two electrode surfaces be expressed as E_{S1} and E_{S2}, an anion as A^-, a cation as C^+, and the electrode/electrolyte interface as //; the electrochemical processes for charging and discharging can be expressed as Eqs. (3.2)–(3.5) [6].

On one electrode (positive):

$$E_{S1} + A^- \overset{\text{Charging}}{\rightarrow} E_{S1}^+ // A^- + e^- \tag{3.2}$$

$$E_{S1}^+ // A^- + e^- \overset{\text{Discharging}}{\rightarrow} E_{S1} + A^- \tag{3.3}$$

On the other electrode (negative):

$$E_{S2} + C^+ + e^- \overset{\text{Charging}}{\rightarrow} E_{S2}^- // C^+ \tag{3.4}$$

$$E_{S2}^- // C^+ \overset{\text{Discharging}}{\rightarrow} E_{S2} + C^+ + e^- \tag{3.5}$$

And the overall charging and discharging process can be expressed as Eqs. (3.6), (3.7)

$$E_{S1} + E_{S2} + C^+ + A^- \overset{\text{Charging}}{\rightarrow} E_{S1}^+ // A^- + E_{S2}^- // C^+ + e^- \tag{3.6}$$

$$E_{S1}^+ // A^- + E_{S2}^- // C^+ \overset{\text{Discharging}}{\rightarrow} E_{S1} + E_{S2} + C^+ + A^- \tag{3.7}$$

3.3 ELECTRODE MATERIAL

3.3.1 Double-Layer Capacitance-Based Materials

3.3.1.1 Activated Carbon (AC) as Electrode Material

Carbon is one of the most abundantly available and structurally diverse materials, and most present-day EDLCs employ porous carbons as the active electrode material. Abundantly available organic materials such as coconut shells, charcoal, nut shells, wood, and food waste [7] are a particularly attractive natural resource for the commercial production of porous carbon materials. The porosity of activated carbon plays a major role in increasing capacitance with an increased surface area. Nonetheless, the carbon structure, including pore shape, surface functional groups, and electrical conductivity, must be considered. Activation treatment is required for improving capacitance by increasing the surface area by opening pores that are closed, clogged, or obstructed [8]. The activation of porous carbons can be achieved in two ways. The first, physical activation, involves treatment at high temperatures from 600°C to 1000°C using steam, CO_2, or air. The second, chemical activation, is carried out between 400°C and 600°C with activating reagents such as KOH, NaOH, H_3PO_4, and $ZnCl_2$ [9].

ACs produced by activation processes have broad pore size distribution, which is divided into micropores (<2 nm), mesopores (2–50 nm), and macropores (>50 nm). The theoretical EDL capacitance ranges from 15 to 25 $\mu F\ cm^{-2}$, but experimentally small EDL capacitance <10 $\mu F\ cm^{-2}$ was obtained with a high surface area up to 3000 $m^2\ g^{-1}$ [10]. This shows that not all pores are effective in charge accumulation [11].

3.3.1.2 Graphene as an Electrode Material

Graphite is becoming an important raw material in many applications, including energy devices, microsensors, superadsorbents, semiconductors, etc. For the conversion of graphite to reduced graphene oxide (rGO), many methods have been used [12,13]. A few green methods have also been implemented, such as the one by Chen [14], without the use of polymer or surfactant. But the use of ammonia in this method is a concern when considered in large amounts. Microwave treatment has also been used for the exfoliation of graphite oxide to rGO [15]. The green method wherein ionic liquid-assisted microwave reduction of GO gave a specific capacitance of 135 Fg^{-1} and was reported to be rapid and facile [16]. Using Gum Arabic and ultrasonification, graphite was exfoliated, but again the concern is that to bring about 100% pure graphene, 100 h of ultrasonification and acid treatment were required [17]. Similarly, many attempts have been made to develop ecofriendly methods to prepare rGO [18,19], which is usually associated with complex processes for the removal of reducing agents. Using sodium carbonate, rGO was efficiently reduced from GO, but it took 4 h for reduction [20].

Graphene has been used as an electrode material in supercapacitors [21]. The graphene honeycomb lattice is composed of two equivalent sublattices of carbon atoms bonded together with σ bonds. A single atomic layer of graphene having a high surface area is the idealistic electrode material, but the removal of heteroatoms and functional groups completely will affect the capacitance. Graphene is known to have a theoretical surface area of 2630 $m^2\ g^{-1}$ and Hantel et al. [22] claimed to have achieved 2687 $m^2\ g^{-1}$ from partially reduced graphene oxide. Fewer layers of graphene showed 1400 $m^2\ g^{-1}$surface area and were porous as prepared using a microwave [23]. Highly porous three-dimensional graphene produced from biomass showed good conductivity and specific capacitance of 231 $F\ g^{-1}$ [24]. Graphene and rGO are slightly different in structure because rGO has more lattice defects and trace amounts of functional groups when compared to graphene. The rGO/polymer binder having a mesoporous rGO was reported for good ionic liquid accessibility and hence showed a specific capacitance of 250 $F\ g^{-1}$ [25]. Even a macroporous low BET surface area significantly adsorbed oil better than micro- or mesoporous graphene sheets. Nevertheless, compared to the surface area of carbon nanofibers and activated carbon, rGO has a lower surface area.

3.3.2 Pseudocapacitance-Based Material

A redox-based electrochemical double-layer capacitor (EDLC) is also called a pseudocapacitor. In a pseudocapacitor, similar to batteries, when a potential is applied a fast reversible reaction occurs, transferring the charge across the electrode/electrolyte interface by overcoming the double-layer barrier. This process is called the Faradaic process. The materials such as conducting polymers and several metal oxides, including RuO_2, MnO_2, and Co_3O_4, undergo redox reactions [6].

Depending upon the type of material used, three types of Faradaic processes mainly occur: (a) reversible adsorption wherein it occurs due to the adsorption of hydrogen on the surface of platinum or gold, (b) redox reactions of transition metal oxides (e.g., RuO_2), and (c) reversible electrochemical doping-dedoping in conducting polymer-based electrodes. The capacitances of the pseudocapacitor are usually higher than the EDLC because the electrochemical redox processes occur both on the surface and in the bulk near the surface of the solid electrode. Conway et al. [10] reported that the capacitance of a pseudocapacitor can be 10–100 times higher than the electrostatic capacitance of an EDLC. Nevertheless, the pseudocapacitor has relatively less power density because redox reactions occur at the electrode and this further increases instability during cycling.

3.3.2.1 *Conducting Polymers (CP)*

Conducting polymers (CP) have many advantages that make them suitable materials for the supercapacitor, including high conductivity in a doped state, a low environmental impact, a low cost, a high voltage window, adjustable redox activity through chemical modification, and high storage capacity/porosity/reversibility [26].

In conducting polymers, the charge can be induced through ion insertion by using suitable doping agents that bring about oxidation or reduction of the conducting polymer. Oxidation on the repeating units of polymer chains brings about a positive charge and is called "p-doped" and negatively charged polymers are generated by reduction and are called "n-doped." The electronic state of p electrons determines the potential of the doping process (Fig. 3.7).

Mechanism

The electrons involved in the nonfaradaic electrical double-layer charging are the itinerant conduction-band electrons of the metal or carbon electrode while the electrons involved in the faradic processes are transferred to or from the valence-electron state (orbital) of the redox cathode or anode reagent. The electrons may, however, arrive in or depart from the conduction support material depending on whether the Fermi level in the electronically conducting support lies below the highest occupied state (HOMO) of the reductant or above the lowest unoccupied state (LUMO) of the oxidant. In pseudocapacitors, the nonfaradaic double-layer charging process is usually accompanied by a faradaic charge transfer. Accordingly, the capacitance C of a super capacitor is given by,

$$C = C_{\mathrm{dl}} + C_{\mathrm{f}} \tag{3.8}$$

C_{dl} = Electrical double-layer capacitance and C_{f} = Pseudocapacitance.

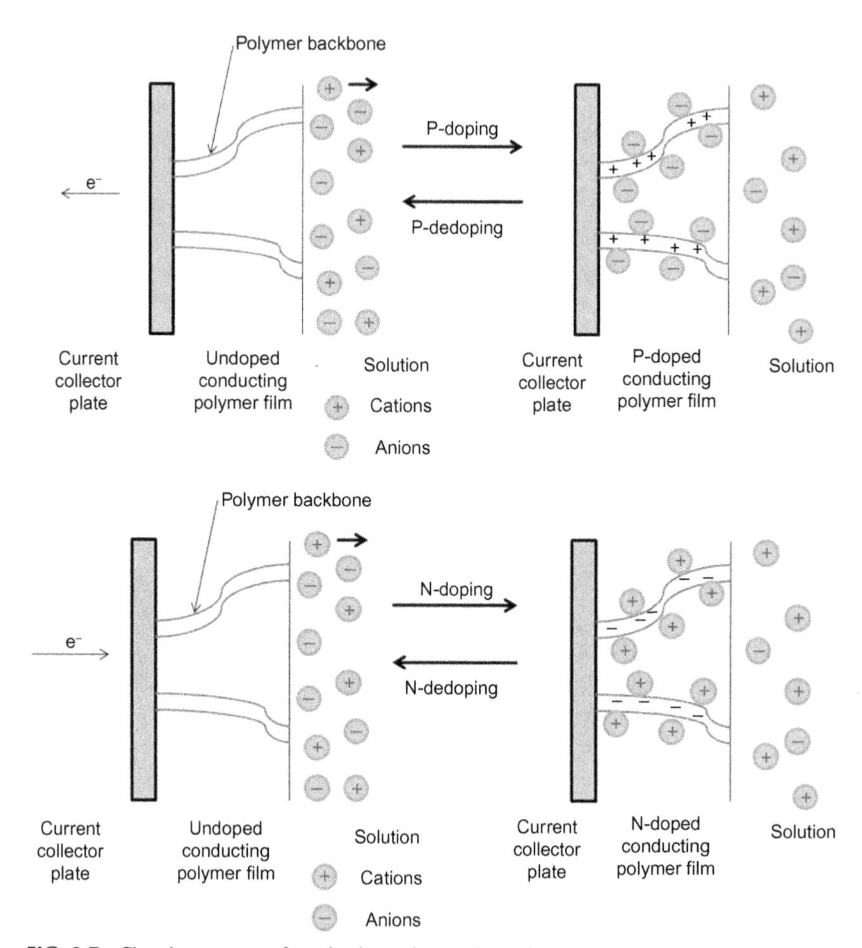

FIG. 3.7 Charging process of conducting polymer electrodes.

The pseudocapacitance is brought about by a surface redox reaction of the type,

$$O_{ad} + n \rightarrow R_{ad} \tag{3.9}$$

where O_{ad}, R_{ad} = adsorbed oxidants and reductants, and n = number of electrons. Some ECs use fast, reversible redox reactions at the surface of active materials, thus defining what is called the pseudocapacitive behavior.

CPs-based supercapacitor systems have three configurations [27]:

Type I: In this type of supercapacitor, both electrodes consist of polymers of the p-dopable type. In the charged supercapacitor, one of the electrode CPs becomes positively charged due to oxidation and the other remains in the uncharged state. When fully charged, a potential difference may go up to

1.5 V and half this potential is available for charging and discharging. During discharging, the charges redistribute between the electrodes.

Type II: In a Type II supercapacitor, two different p-dopable polymers such as polypyrrole/polythiophene with different ranges of potentials for oxidation and reduction are used.

Type III: In a Type III supercapacitor, polymers that can be both p- and n-dopable are used. In this supercapacitor, the discharging half cycle of the p-doped electrode is worked against the discharge half cycle of the n-doped electrode. When a nonaqueous electrolyte is used, the potential developed will have a wider range up to 3.5 V and there is an increase in energy density. Significant work is required in developing Type III supercapacitors, including advancement in polymer technology, material design, and the capability of storing energy while operating at high voltages.

3.3.2.2 *Metal Oxides*

Metal oxides have a tendency to undergo reversible oxidation and reduction processes over a wide range of potentials. Usually, metal oxides operate at higher potentials (1–4 V) when compared with a redox couple such as $Fe(CN)_6{}^{4-}/Fe(CN)_6{}^{3-}$ (0.2 V). The potential is obtained not only from pseudocapacitance but from the double-layer capacitance. This makes metal oxides an important commercially applicable material that can provide higher energy density for supercapacitors than conventional carbon materials. The lattice is quite stable during charging and discharging at various current densities, hence they have better electrochemical stability than conducting polymer materials. Some examples such as ruthenium oxide (RuO_2), manganese oxide, cobalt oxide, nickel oxide, and vanadium oxide are extensively used for energy devices.

Some characteristic requirements for metal oxides for supercapacitor applications are:

(a) The metal oxide should be electronically conductive.
(b) The metal can exist in two or more oxidation states that coexist over a continuous range with no phase changes involving irreversible modifications of a three-dimensional structure for electronic conductivity.
(c) The protons can freely intercalate into the oxide lattice on reduction (and out of the lattice on oxidation), allowing facile interconversion of $O^{2-} \leftrightarrow OH^-$. The porosity helps in the diffusion of protons inside the electrode layers.
(d) A hydrous structure to support proton diffusion, nanosized.
(e) High packing density to increase volumetric energy density and capacitance.
(f) Ability to withstand charge-discharge cycles.

Some examples: ruthenium oxide (RuO_2), manganese oxide, cobalt oxide, nickel oxide, and vanadium oxide.

RuO_2/Polymer-Based Electrodes

Metal oxides doped in polymer electrolytes have several roles in contribution supercapacitors:

(a) Aggregation of metal oxide particles is avoided by electrostatic stabilization and steric mechanisms of polymer electrolytes.
(b) Metal oxide particles are uniformly distributed on the polymer electrolyte.
(c) Polymer electrolytes enhancing the active surface area of the metal oxide particles.
(d) Providing a pathway for proton conduction along with metal oxide.
(e) Increasing the effectiveness of the metal oxide by improving the adhesion to the electrode.

A few drawbacks such as a large volume of polymer electrolyte along with the metal oxide cause an increase in weight, which is unsuitable for light and thin energy devices. During charging-discharging of these composites, the polymer electrolytes expand and contract due to the vibration and motion of the metal oxide through the polymer matrix. This is another critical problem in the long run for the energy device.

RuO_2/polymer composites are widely studied and applied in various energy devices. For example, poly(3,4-ethylenedioxythiophene)-doped poly(styrene sulfonic acid) doped with hydrous RuO_2 particles exhibited a high specific capacitance of $653\ F\ g^{-1}$. A nanostructured polypyrrole/RuO_2 composite reported by Zang et al. [28] achieved a specific capacitance of $302\ F\ g^{-1}$. A RuO_2/poly(3,4-ethylenedioxythiophene) nanotube composite showed fast charging/discharging capability along with high specific capacitance of $1217\ F\ g^{-1}$. This supercapacitor exhibited a high power density of $20\ kW\ kg^{-1}$ and energy density ($28\ Wh\ kg^{-1}$) up to 80%. This enhanced property was interpreted to the nano hollow tubular structures of metal oxide, which readily allowed ions to penetrate and access the internal surfaces, thereby decreasing the hindrance of ion transport through the polymer matrix.

Measurements

Dielectric Constants Complex impedance data Z^* can be represented by its real, Z_R, and imaginary, Z_I, parts by the relation:

$$Z^* = Z_R + jZ_I \tag{3.10}$$

The relationships between complex impedance, admittance, permittivity, and electrical modulus can be found elsewhere [29]. The equations for the dielectric constant, ε_R, the dielectric loss, ε_I, the real electrical modulus M_R, and the imaginary electrical modulus M_I can be shown as

$$\varepsilon_R = \frac{Z_I}{\omega C_o (Z_R^2 + Z_I^2)} \tag{3.11}$$

$$\varepsilon_I = \frac{Z_R}{\omega C_o (Z_R^2 + Z_I^2)} \tag{3.12}$$

$$M_R = \frac{\varepsilon_R}{(\varepsilon_R^2 + \varepsilon_I^2)} \tag{3.13}$$

$$M_I = \frac{\varepsilon_I}{(\varepsilon_R^2 + \varepsilon_I^2)} \tag{3.14}$$

Here $C_0 = \varepsilon_0 A/t$ and ε_0 is the permittivity of the free space, A is the electrolyte-electrode contact area and t is the thickness of the sample and $\omega = 2\pi f$, f being the frequency in Hz.

Cyclic Voltammetry Cyclic voltammetry involves potential modulation at a rate $\pm dV/dt$ between two potential limits, ΔV (V_1 and V_2). The resultant is a potential and time-dependent current represented in Eq. (3.15) and a waveform observed in Fig. 3.8.

$$i = C(dV/dt) \tag{3.15}$$

Depending upon the information required, the experiment can be performed for a number of cycles. The resulting response is plotted as current (A) versus potential (V), called cyclic voltammogram. This is a convenient method for characterization of both double-layer and pseudocapacitance behavior of supercapacitors. The pseudocapacitance behavior shows an anodic peak and a cathodic peak in mirror-image symmetry during a reversible process (Fig. 3.9). The electrode materials such as conducting polymers and metal oxide usually exhibit pseudocapacitance behavior because their oxidation and reduction potential lies in the study range.

The electrochemical double-layer CV response shows rectangular behavior without any peaks for the ideal supercapacitor because no redox reaction occurs

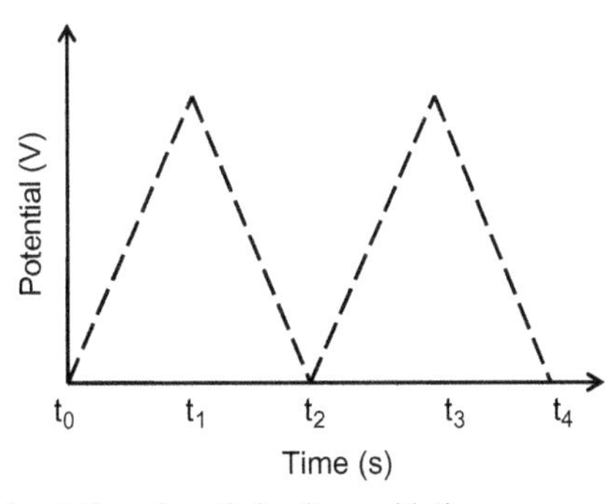

FIG. 3.8 Waveform of variation of applied voltage with time.

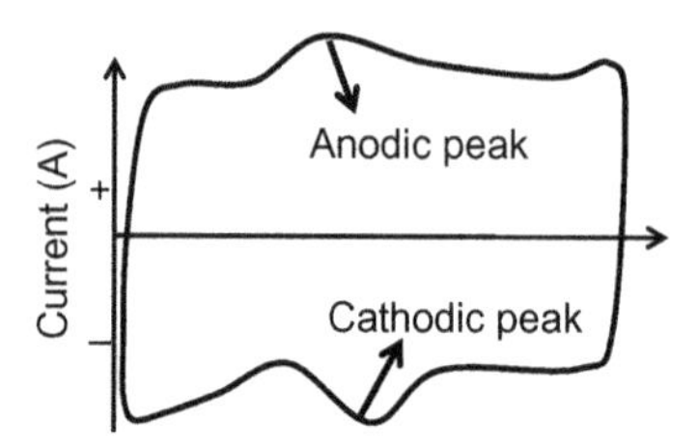

FIG. 3.9 CV response pseudocapacitance behavior of supercapacitor.

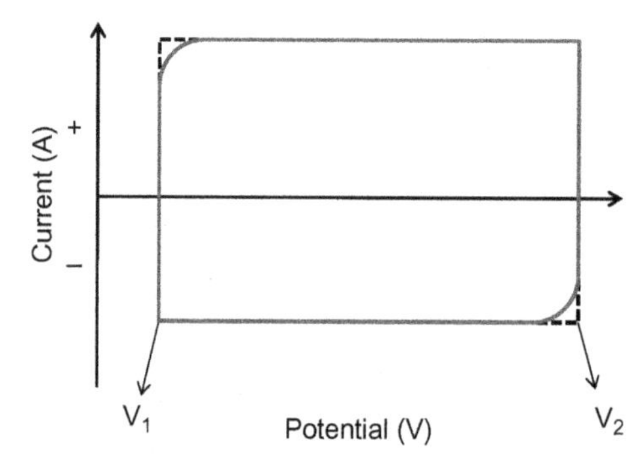

FIG. 3.10 CV response between potential difference V_1 and V_2 of ideal double-layer capacitor (dotted line) and nonideal double-layer capacitor.

at the electrode-electrolyte interface (Fig. 3.10). The electrode materials such as activated carbon, graphene, and materials that are inert to the electrolyte do not exhibit the redox peaks. Due to the presence of electrochemical series resistance (ESR), the property of the material used slightly deviates the CV response from the ideal rectangular behavior.

The specific capacitance values of the supercapacitors have been calculated using the equation:

$$C_s = \frac{2\Delta I}{\Delta V \times m} \tag{3.16}$$

where I is the average current, ΔV is the potential window, v is the scan rate, and m is the mass of materials at each electrode. The factor of 2 is used because the series capacitance is formed in a two-electrode system.

Galvanostatic Charge-Discharge Studies (GCD)

The stability of the materials in the supercapacitor is usually tested by using the GCD method. Here a fixed current is applied within a fixed range of cell potential and the resulting variation of the potential response is measured as a function of time. In GCD studies, the curves with triangular symmetrical distribution indicate good capacitive properties of the supercapacitor. The specific capacitance was derived from the charge-discharge curve according to the following equation:

$$C_s = \frac{I \times \Delta t}{\Delta V \times m} \tag{3.17}$$

where I is the applied discharge current, Δt is the discharged time after IR drop, ΔV is the discharge potential window after IR drop, and m is the mass of the single electrode materials, respectively.

In the supercapacitor, it becomes possible to store enough energy for many practical purposes. Hence, energy density (E) is an important value to evaluate the device. It may be pointed out that because voltage decreases linearly with the state of charge, not all the stored energy can generally be used. The factors influencing energy density are the dielectric constant, surface area, etc., but the factors determining power density are much more complex. The power density (P) is inversely proportional to the resistance of the device; R. R is the equivalent series resistance (ESR), which is comprised of the electrode resistance, the electrolyte resistance, and resistance due to the diffusion of ions in the electrode porosity. The energy density, power density, and ESR were calculated from the following equations:

$$E = \frac{C \times \Delta V^2}{2} \times \frac{1000}{3600} \tag{3.18}$$

$$P = \frac{I \times \Delta V}{2 \times m} \times 1000 \tag{3.19}$$

$$ESR = \frac{iR_{\text{drop}}}{2 \times I} \tag{3.20}$$

Cyclic durability is an important electrochemical performance of the supercapacitor for its practical application. Supercapacitors were subjected to many charge-discharge cycles under constant current density. The Columbic efficiency (η) was used to measure the efficiency of the supercapacitor by subjecting it to several charge-discharge cycles; it was calculated using the formula:

$$\eta = \left(\frac{\text{Discharge current}}{\text{Charge current}}\right) \times 100 \tag{3.21}$$

3.3.3 Advantages

(1) High power density. As the name given as electrochemical capacitor due to storage of electrical charges as double-layer rather than exchange of charge between the bulk of the device and entire surface of electrode, it is clear that EDLC shows a much higher power delivery (in ~10,000 W kg^{-1}) when compared to lithium ion batteries (~150 W kg^{-1}). Rapid charging and discharging rates lead to a high power density in the EDLC.

(2) Long life expectancy. As compared to batteries, the electrochemical capacitor does not need maintenance because the small charge transfer and phase variations will not make it irreversible as compared with redox reactions involved in a battery. Hence, charge/discharge can last for several million cycles.
(3) Long shelf life. EDLC has a lower self-discharge rate and can stay operational for many years without use. Most rechargeable batteries that are unused for months, however, will degrade and become useless due to self-discharge and corrosion.
(4) Wide range of operating temperatures. An EDLC can function effectively at extremely high and low temperatures. This ability is advantageous for military applications.
(5) Environmental friendliness.
(6) Safety. EDLC is much safer than lithium ion batteries.

3.3.4 Challenges for EDLC

(1) Low energy density. To improve the low energy density in EDLC requires large construction, thereby increasing the cost. So it is a challenge to design and use material for high energy density that nears that of the battery ($\sim$30 Wh kg^{-1}).
(2) High cost. The costs of raw materials such as like high surface area carbon, metal oxides, and electrochemical capacitor manufacturing continue to be major challenges for EDLC commercialization.
(3) Industrial standards for commercialization. Fabrication is a key for the electrochemical supercapacitor with a low cost and efficient materials. Several studies have been done in last decade, but still carbon/carbon-based EDLC with a capacitance of 50–5000 F are commercially available due to low cost than metal oxide based EDLC.

3.3.5 Applications of EDLC

EDLCs are playing a very significant role in applications such as electric vehicles, electric hybrid vehicles, digital communication devices, digital cameras, mobile phones, electrical tools, pulse laser techniques, uninterruptible power supplies, and storage of the energy generated by solar cells. Hybridizing EDLC with batteries is under research; this could certainly improve performance in hybrid electric vehicles, including power acceleration, braking energy recovery, excellent cold weather starting, and increased battery life. Thus, EDLC has the potential to play an important role in complementing or replacing batteries in the energy conversion and storage fields. The main market targeted by EDLC manufacturers in the coming decades may be transportation, including hybrid electric vehicles and metro trains.

3.4 ELECTROLYTES

The investigations discussed in this book are based on new concepts in the section of biopolymer electrolytes prepared by blending, gelation, and solution casting techniques for supercapacitor applications. The work involves components from various area of research and hence is interdisciplinary in nature. Polymer electrolytes are the most widely studied solid type of electrolyte. The advantages of using polymer electrolytes as discussed earlier in Chapter 1 serves as an energy storage electrolyte as well as a separator.

3.4.1 Blend Biopolymer (BBPE) Electrolytes as a Supercapacitor

Kadir et al. [30] prepared a blend polymer electrolyte doped with NH_4NO_3 for understanding the interaction between the host polymer and doped salt. Chitosan (CS) is an abundant natural polymer with excellent film-forming property, gelling, biodegradability, and resistance toward chemicals. Chitosan is compatible with many natural as well as synthetic polymers, which thereby enhances the property compared to individual polymers. Chitosan has been used as a blend polymer electrolyte with combination acids, alkalis, and lithium salts. Plasticized CS/PEO-doped ammonium nitrate electrolyte films obtained the highest conductivity of 2×10^{-3} S cm^{-1} and the fabricated supercapacitor showed 134 mF g^{-1} [31]. Conductivities of the order of 10^{-6} S cm^{-1} at room temperature were reported for chitosan/poly(ethylene oxide) PEO blends with LiTFSI salt and for the complex formed by chitosan, poly(aminopropylsiloxane) (pAPS), and $LiClO_4$ [32].

High molecular weight poly(ethyleneglycol) (PEG) has a good film-forming property for which it is used in blending with other polymers. PVA-PEG-$Mg(NO_3)_2$ polymer blend electrolytes showed 9.6×10^{-5} S cm^{-1} and their dielectric studies showed that the polymers are polar in nature [33]. Ductility was improved by modifying the chitosan by preparing blends and grafting with PEG [34]. PEG blending with PVC helped to enhance the dissociation of TiO_2 and increase the conductivity [35]. Hence the PEG is a plasticizing property and provides channels for the conduction of lithium ions or cations because of the presence of the OH group in the polymer chain.

Humidified poly(styrenesulfonic acid) (PSSA) has high proton conductivity and for a short period it is mainly used as a polymer electrolyte membrane in fuel cells [36]. Further, Acar et al. [37] prepared poly(2, 5-benzimidazole)/PSSA blend membranes that have thermomechanical stability and illustrated sufficient proton conductivities under low humidity conditions.

A Case Study on Lithium Salts-Doped Plasticized Chitosan and Poly(Ethyleneglycol) Blend as a Biodegradable Polymer Electrolyte for a Supercapacitor

Experimental

Materials Preparation Chitosan (medium molecular weight from crab shells) and poly(ethylene glycol) (PEG) (number average molecular weight 35,000) were purchased from Aldrich. The plasticizers, ethylene carbonate (EC), and propylene carbonate (PC) from Sigma-Aldrich were used as such without further purification. Lithium perchlorate (Aldrich) was dried at 393 K and kept under vacuum for 48 h before use. A 1% (w/v) chitosan solution was prepared by dissolving it in 1% acetic acid and similarly 1% (w/v) PEG in double-distilled water separately. These polymer solutions were used as stock solutions. CS/PEG blend solutions were prepared by mixing the appropriate amount of these stock solutions with salt and plasticizer, which were subsequently employed as indicated in Table 3.1.

TABLE 3.1 Film Content of the CS/PEG Blend Electrolyte

Blend Ratio of Polymers (CS:PEG)	Ratio of Doped Salt + Plasticizers				
	Dopant Salt	*Propylene Carbonate*	*Ethylene Carbonate*	**Total wt.%**	**Sample Number**
90:10 (70 wt.%)	0.8	4.2	5.0	100	S1
80:20 (70 wt.%)	0.8	4.2	5.0	100	S2
70:30 (70 wt.%)	0.8	4.2	5.0	100	S3
60:40 (70 wt.%)	0.8	4.2	5.0	100	S4
70:30 (100 wt.%)	0.0	0	0	100	S5
70:30 (99.5 wt.%)	0.5	0	0	100	S6
70:30 (99 wt.%)	1.0	0	0	100	S7
70:30 (98.5 wt.%)	1.5	0	0	100	S8
70:30 (98 wt.%)	2.0	0	0	100	S9 (control)

polymer electrolytes were dried initially at room temperature and were then kept in a vacuum oven at 333 K for 48 h to complete dryness before subjecting to various studies.

For $Li_2B_4O_7$, the sample was prepared by mixing the appropriate amounts of salt and polymer blends (CS/PEG) as follows (a) 90/10 (CS/PEG 70 wt.%) + (0.5:4.5:5.0) $Li_2B_4O_7$:PC:EC, 30 wt.%), (b) 80/20 (CS/PEG 70 wt.%) + (0.5:4.5:5.0) $Li_2B_4O_7$:PC:EC, 30 wt.%), and (c) 70/30(CS/PEG 70 wt.%) + (0.5:4.5:5.0) $Li_2B_4O_7$:PC:EC, 30 wt.%).

Electrochemical Studies Samples were cut into a 1 cm^2 area and placed between two square copper electrodes (length 1 cm) fitted with copper wires. The whole setup was held tightly with a plastic clamp. The bulk ionic conductivities (σ) and dielectric properties of the blends were determined from the electrochemical impedance spectra (EIS) in the frequency range between 1 and 100 MHz using a small amplitude AC signal of 10 mV. Experiments were carried out at a temperature range of 298–343 K using a proportional-integral-derivative controlled oven from SES Instruments Pvt. Ltd. Potentiostatic polarization experiments were done with an applied voltage of 10 mV for 1 h; impedance before and after was measured in the frequency range of 1–100 MHz.

Fabrication of a Symmetrical Supercapacitor Cell The electrode material for supercapacitor fabrication was prepared using activated carbon (AC) derived from areca fibers having a surface area 250 $m^2 g^{-1}$. AC was coated on a stainless steel electrode using poly(vinylidene fluoride) as the binder. The supercapacitor cell was constructed using $LiClO_4$-doped plasticized CS/PEG blend film sandwiched between two AC-coated electrodes. Electrochemical characterization was carried out by cyclic voltammetry (CV), EIS, and galvanostatic charge-discharge studies. All the electrochemical studies were carried out using a BioLogic SP-150 electrochemical system.

Results and Discussion

Ionic Conductivity Studies The temperature dependence of conductivity for the chitosan/PEG blend ratios is shown in Fig. 3.11.

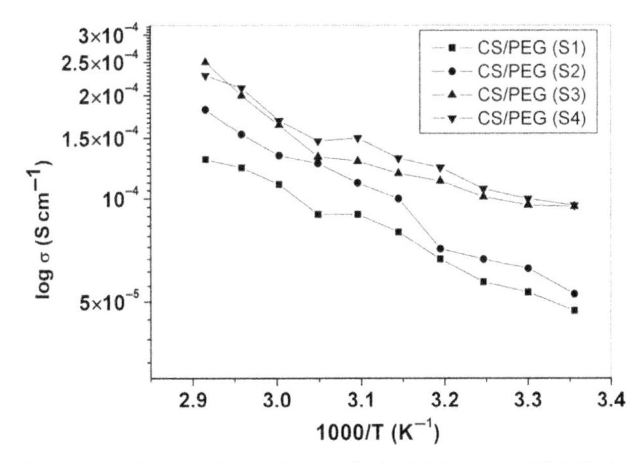

FIG. 3.11 Variations of conductivity of plasticized $LiClO_4$-doped CS/PEG blend films of different compositions with temperature.

The plot shows that as temperature increases, conductivity increases. Films containing little PEG have the lowest conductivities. On the other hand, solid films obtained from 40% above content of PEG (S4) in the CS/PEG blend ratio segregated into fragments while being taken from the petri dish. At the same time, for this sample S4, the conductivity was found to be higher at all temperatures although it could not provide free-standing film due to the poor mechanical property of the PEG rich phase. Films with an intermediate CS/PEG blend ratio of 70:30 possess good free-standing properties with reasonable conductivity. Thus, to obtain a good freestanding film, it is necessary to add chitosan. Therefore, S3 (70:30) was selected as the biodegradable blend polymer electrolyte based on its conductivity and freestanding nature for further studies.

The maximum conductivity for S3 was found to be 1.1×10^{-4} S cm^{-1} at room temperature and 2.7×10^{-4} S cm^{-1} at 343 K. The ionic conductivity for a sample (S9) without plasticizer was 8.7×10^{-6} S cm^{-1}. On the addition of plasticizer, the conductivity value increased to $\sim 10^{-3}$ S cm^{-1} and continued to increase with increasing plasticizer as well as salt content. It can be inferred that the salt is responsible for the conductivity of the CS/PEG blend films in the plasticizer medium. From these results, it is suggested that in the electrolyte films containing both chitosan and PEG, the chitosan-rich phase acts as mechanical support and the plasticizer-rich phase acts as a tunnel for ionic transport. Although the chitosan-rich phase also interacts with the salt ions, it is a solid-like medium in which it is difficult for ions to penetrate this phase. Thus transport of ions must occur through a convoluted path within the chitosan and also via indirect motion along the plasticizer-phase, which may be responsible for low conductivity. PEG-rich film presents a homogeneous medium through which ions can drift across, as reflected in the higher conductivity of such films. In the $Li_2B_4O_7$-doped blend polymer electrolyte, the maximum conductivity was 3.19×10^{-5} S cm^{-1} at 343 K and 7.9×10^{-5} S cm^{-1} at room temperature (Fig. 3.12).

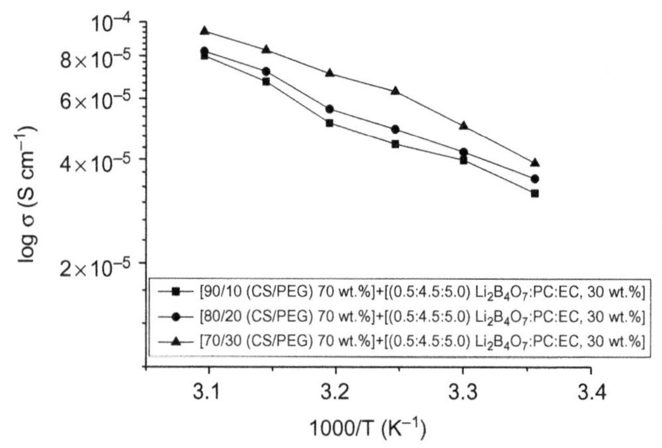

FIG. 3.12 Variations of conductivity of plasticized $Li_2B_4O_7$-doped CS/PEG blend films of different compositions with temperature.

Bruce and Vincent established a potentiostatic polarization method for ideal solid polymer electrolytes [38]. Applying a small constant potential on an electrolyte between electrodes leads to a decrease of the initial current value until a steady-state value is reached. If no redox reaction occurs with the anions, the anion current will vanish in the steady state and the total current will be caused by the cations. In this case, for small polarization potentials (10 mV), the transference number for the cation t_+ is calculated using the formula reported in the literature [39]. The determined lithium ion transference number is ~0.72 for the S3 sample and implies that the system is ionic motion, particularly in this blend polymer electrolyte as Li^+ has the foremost role in ionic conductivity.

The activation energy, E_A, was calculated from the slope of the log σ versus $10^3/T$ graph (Fig. 3.12). This behavior is based on the Arrhenius rule Fig. 3.13 which shows activation energy and room temperature conductivity of various samples (S1, S2, S3, and S4) as a function of chitosan in the blend ratio.

The material with the highest conductivity shows the minimum activation energy, as expected. The values of the conductivity and activation energies as a function of chitosan concentration show that the activation energy for the conduction decreased gradually with the increase in chitosan concentration. This may be due to an increase in rigidity in the blend matrix on the addition of chitosan, thereby decreasing the dissociation of the dopant salt. For chitosan-$LiCF_3SO_3$, the activation energy $E_A = 0.36$ eV obtained with the impedance technique is close to our results. Similarly, when compared with the ionic conductivity of 4.0×10^{-5} S cm^{-1} as reported by Osman et al. [40] for chitosan, the acetate is lower than that of our system.

Among the various samples, the bulk ionic conductivities (σ) of the sample (S3) have been determined from the complex impedance spectra (Fig. 3.14).

The bulk resistance was calculated from the high frequency intercept on the real impedance axis of the Nyquist plot as a function of temperature. It is noticeable

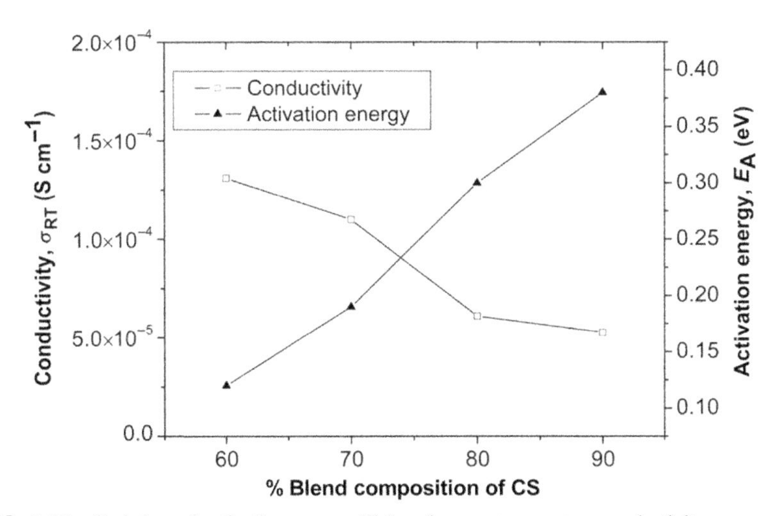

FIG. 3.13 Variation of activation energy (E_A) and room temperature conductivity, σ_{RT}, as a function of chitosan for various samples.

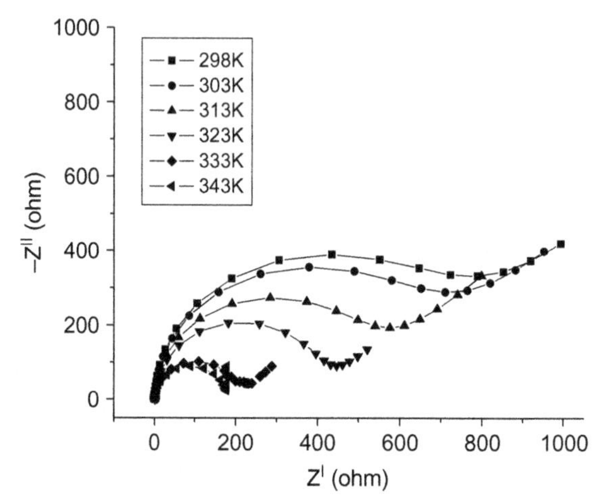

FIG. 3.14 Nyquist impedance plots of the biodegradable polymer electrolyte sample (S3) at different temperatures.

from Fig. 3.14 that the resistance of the films decreases with an increase in temperature. Due to comparatively high resistance at room temperature (298 K), the semicircle is not complete, that is, the R_b value is 820 Ω and the semicircle becomes more clear and complete ($R_b = 190\,\Omega$) with an increase in the diffusion of ions in the electrolyte at 343 K. As temperature increases, the polymer chains acquire faster internal modes in which bond rotations produce segmental motion, favoring hopping inter- and intrachain ion movements and, accordingly, the increase in the conductivity of the polymer electrolyte [41]. The bulk resistance of $Li_2B_4O_7$-doped CS/PEG blend films (Fig. 3.15) is greater than the $LiClO_4$-doped CS/PEG blend films due to a better interaction between salt and polymer, leading to a decrease in ionic conductivity. Hence, $LiClO_4$-doped CS/PEG blend films were chosen for further studies.

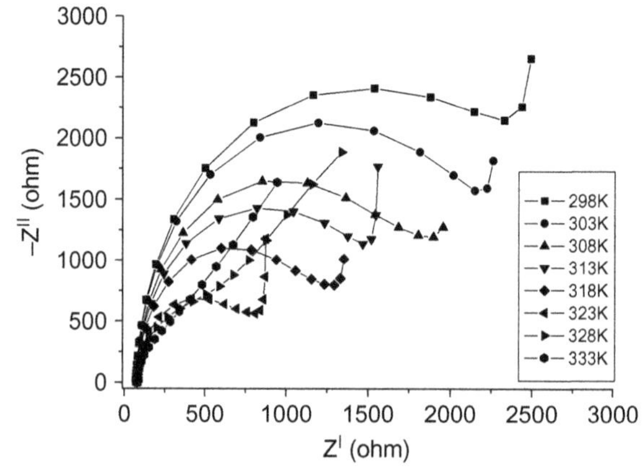

FIG. 3.15 Nyquist impedance plots of the $Li_2B_4O_7$-doped CS/PEG blend films at different temperatures.

Dielectric Studies The conductivity behavior of the CS/PEG blend electrolytes can be understood from the dielectric studies [42]. The dielectric constant is a measure of stored charge. The variation of dielectric constant and dielectric loss as a function of temperature for samples with and without plasticizer is shown in Figs. 3.16 and 3.17, respectively. There are no appreciable relaxation peaks observed in the frequency range used in the study. Both dielectric constant and dielectric loss increase sharply at low frequencies, indicating that electrode polarization and

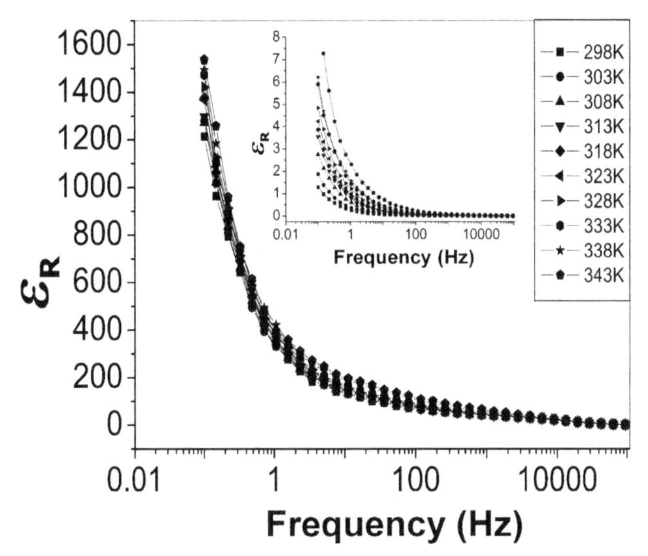

FIG. 3.16 Plots of the dielectric constant versus frequency at different temperatures (inset: without plasticizer (S9)).

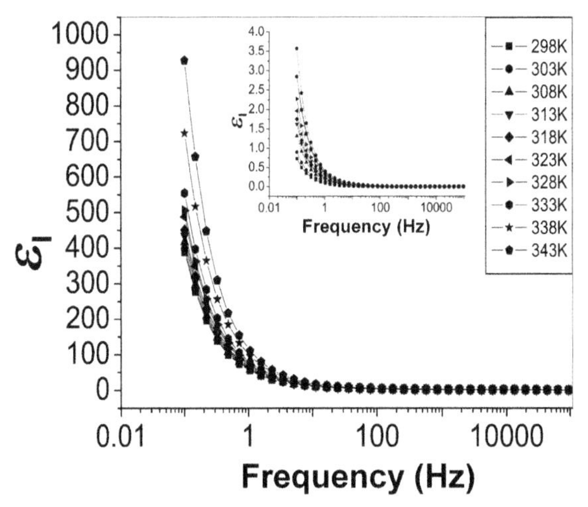

FIG. 3.17 Plots of the dielectric loss versus frequency at different temperatures (inset: without plasticizer (S9)).

space charge effects have occurred, confirming non-Debye dependence. On the other hand, at high frequencies, periodic reversal of the electric field occurs so fast that there is no excess ion diffusion in the direction of the field. Polarization due to charge accumulation decreases, leading to the observed decrease in dielectric constant and dielectric loss [42,43]. The dielectric constant and dielectric loss increase at higher temperatures.

In Fig. 3.16, the inset figure shows ε_R values for the film without plasticizer. ε_R has high values at low frequencies. This implies that in this sample the mobile ions tend to get accumulated at low frequencies, but the ε_R value of the $LiClO_4$-doped plasticized CS/PEG blend film is much higher (up to 200 times) than that without plasticizer. With an increase in plasticizer content, the free volume within the blend polymer matrix increases, also increasing the ability of the biodegradable polymer to dissolve salt. Therefore the number of ions in the sample increases and thus conductivity increases [44].

In the dielectric loss-frequency spectrum, no significant relaxation peaks have been observed in Fig. 3.17. It is inferred that residual water does not contribute toward conductivitys enhancement. The variations of real (M_R) and imaginary (M_I) parts of the electrical modulus for the plasticized CS/PEG blend electrolyte and that without plasticizer (inset) are depicted in Figs. 3.18 and 3.19, respectively. Both M_R and M_I show an increase at the higher frequency end and exhibit a long tail feature at the low frequency end. This indicates that the material is very capacitive in nature [45]

Thermal Behaviors A DSC curve for (S5, S7–S9) nonplasticized CS/PEG blend films is presented in Fig. 3.20. In Fig. 3.20, obtained by secondary scanning, the samples (S5, S7–S9) shows two peaks as a function of increasing $LiClO_4$ dopant in polymer blend films. The presence of two peaks is apparent to the partial immiscible

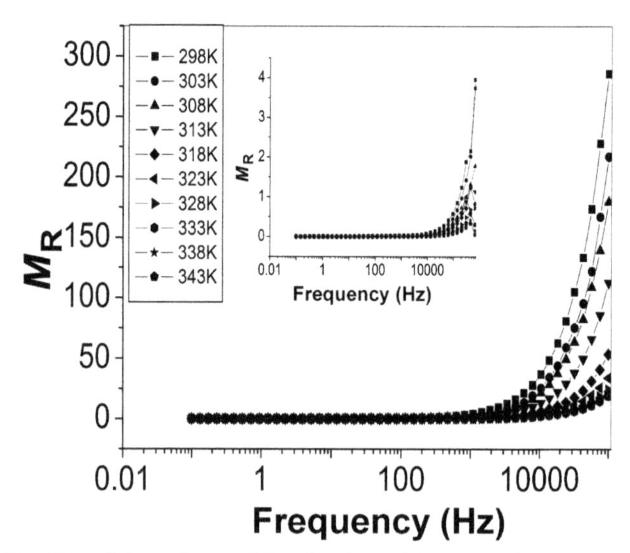

FIG. 3.18 Plots of the real part of the electric modulus versus frequency at different temperatures (inset: without plasticizer (S9)).

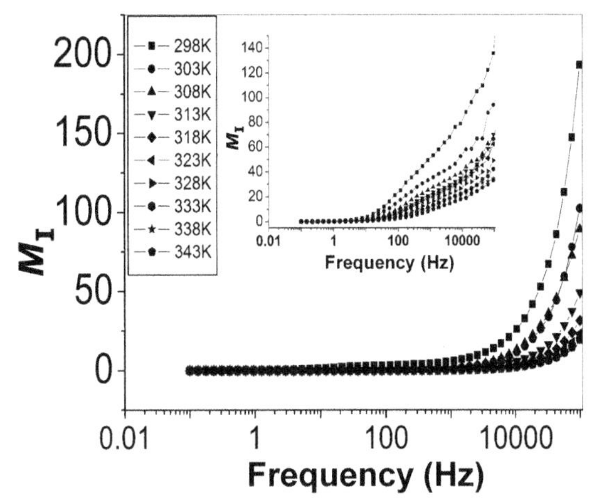

FIG. 3.19 Plots of the imaginary part of the electric modulus versus frequency at different temperatures (inset: without plasticizer (S9)).

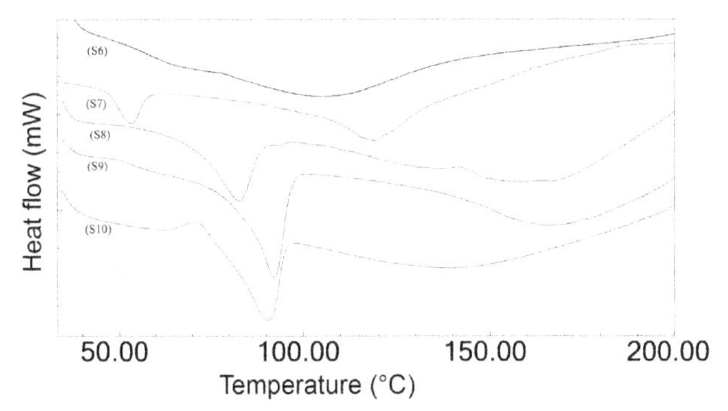

FIG. 3.20 DSC thermograms of various samples containing $LiClO_4$-doped CS/PEG blend polymer electrolyte.

characteristic of the blends. The first peak can be assigned to T_m of the PEG in the blend films. The melt peak increased gradually and became more obvious from 55°C to 100°C with an increase in Li salt content. The second broad peak corresponding to chitosan is read around 125–166°C. With more content of Li salt, the peak became broader and shifted to the higher temperature region. It is probably attributed to solvent because the films contain acetic acid, residual water, bound moisture, etc. [46]. The main reason for the broad peak may be that, being a natural polymer, some properties such as crystallinity, molecular weight, and deacetylation degree can present wide variations according to the source and/or method of extraction and will influence the T_g. In our case, there is no obviously readable T_g for chitosan in the DSC curve. This increase in both the peaks of the blend polymers with increasing salt concentration can be attributed to three interrelated phenomena: (a) an increase in macromolecular rigidity of the amorphous phase because of high salt concentration that acts as a reticulate agent in the polymeric matrix, (b) the presence of crystalline regions at various degrees of perfection within the complexed samples, favoring the appearance of ion pairs, and (c) the appearance of a crystalline phase formed by the salt-polymer complex [47].

For plasticized $LiClO_4$-doped CS/PEG blend film (S3), a single broad peak is observed at 110°C, indicating no phase separation. This also implies that the addition of plasticizer dissociates the dopant salt, which in turn increases the entropy of mixing [47]. This increase in the entropy of mixing along with increasing segmental movements of PEG in the plasticizer phase leads to a reduction in the crystallinity and an enhancement in the flexibility of the molecular system, thereby increasing the ionic conductivity. Consequently, these results agree with ionic conductivity measurements for the plasticized and nonplasticized samples.

Fig. 3.21 shows DSC thermograms of the following samples doped with $Li_2B_4O_7$, (a) 90/10 (CS/PEG 70 wt.%) + (0.5:4.5:5.0) $Li_2B_4O_7$:PC:EC, 30 wt.%), (b) 80/20 (CS/PEG 70 wt.%) + (0.5:4.5:5.0) $Li_2B_4O_7$:PC:EC, 30 wt.%), and (c) 70/30(CS/PEG 70 wt.%) + (0.5:4.5:5.0) $Li_2B_4O_7$:PC:EC, 30 wt.%) compared with a pure starch thermogram. The shift in peak due to PEG in all the blends decreased (72–64°C) with an increase in PEG concentration. This indicates the importance of studying the miscibility of the blend interaction.

Supercapacitor Studies The supercapacitor cell was constructed with the $LiClO_4$-doped plasticized CS/PEG blend sample S3 due to its higher ionic conductivity. Cyclic voltammetry (CV) responses for the carbon-carbon symmetrical supercapacitor at various sweep rates are shown in Fig. 3.22. The behaviors observed are characteristic of double-layer capacitive features. The carbon-carbon supercapacitor fabricated using this blend polymer electrolyte showed a maximum specific capacitance of $47\,F\,g^{-1}$ at a scan rate of $10\,mV\,s^{-1}$.

The AC impedance response (Nyquist plot) of the carbon-carbon supercapacitor is shown in Fig. 3.23. The plot shows a semicircle of a large radius at the high-frequency range and a straight line in the low-frequency region. The large radius exhibiting large resistance is due to the increase of the electrode thickness; the ions

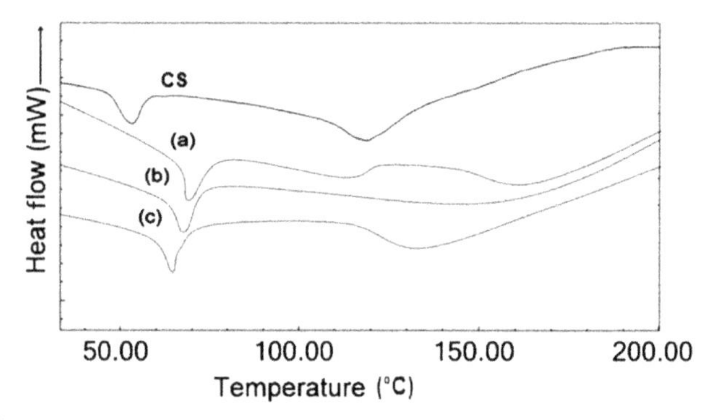

FIG. 3.21 DSC thermograms of various samples containing $Li_2B_4O_7$-doped CS/PEG blend polymer electrolyte.

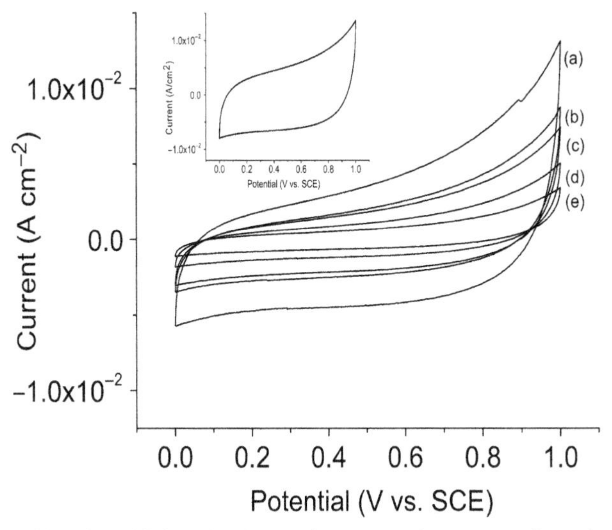

FIG. 3.22 CVs of the fabricated carbon–carbon symmetric supercapacitor using $LiClO_4$-doped (S4) CS/PEG blend electrolyte at scan rates. (a) $50\,mV\,s^{-1}$, (b) $25\,mV\,s^{-1}$, (c) $20\,mV\,s^{-1}$, (d) $10\,mV\,s^{-1}$, and (e) $5\,mV\,s^{-1}$ (inset: CV of activated carbon).

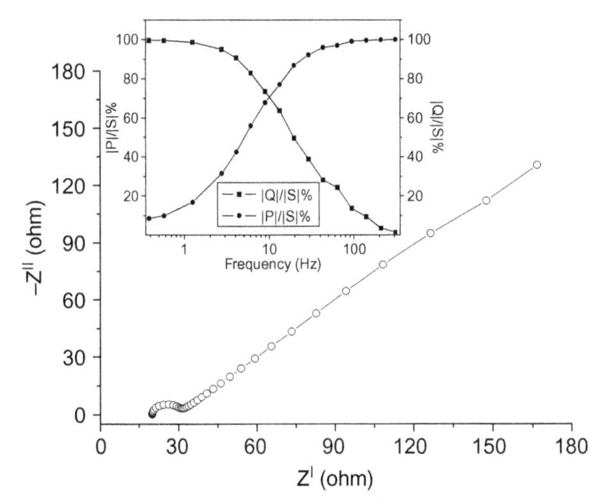

FIG. 3.23 AC impedance plot for a fabricated carbon-carbon symmetric supercapacitor using biodegradable polymer electrolyte (inset: plots of normalized reactive power $|Q|/|S|\%$ and active power $|P|/|S|\%$ versus frequency (Hz) of a supercapacitor).

from the electrolyte have not reached the whole electrode porosity [29]. Using this resistance at high frequency, the value of C_{dl}, the double-layer capacitance, has been determined from the high-frequency region of the impedance spectrum. The C_{dl} value obtained was $6\,mF\,cm^{-2}$. The capacitance value increases at low frequencies because of a large number of ionic movements. The semicircle results from the parallel combination of resistance and capacitance and the linear region is because of Warburg impedance. In the low-frequency region, the straight line part leans more toward an imaginary axis at lower frequencies, which indicates good capacitive behavior.

Using a normalized reactive power |Q|/|S|% and active power |P|/|S|% versus frequency plot for the 1 cm^2 cell, the time constant of the fabricated supercapacitor has been calculated and is shown in Fig. 3.23 (inset). The theoretical details of this plotting technique have been taken from the literature [29].

The impedance $Z(\omega)$ data obtained from the supercapacitor can be written under its complex form:

$$Z(\omega) = Z'(\omega) + j \times Z''(\omega) \tag{3.22}$$

where $Z'(\omega)$ and $Z''(\omega)$ are the real part and the imaginary part of the impedance, respectively. j the complex number (−1) and ω the angular frequency. To describe the supercapacitor by using capacitance that is a function of the pulsation, ω is noted as $C(\omega)$ and can be stated as

$$C(\omega) = C'(\omega) - j \times C''(\omega) \tag{3.23}$$

where C' is the real capacitance of C'' is the imaginary capacitance of the super capacitor varying with frequency and can be defined as Eqs. (3.24), (3.25)

$$C'(\omega) = -Z''(\omega)/\omega \times |Z(\omega)|^2 \tag{3.24}$$

$$C''(\omega) = Z'(\omega)/\omega \times |Z(\omega)|^2 \tag{3.25}$$

where $C'(\omega)$ is the real part of the capacitance $C(\omega)$. The low-frequency value of $C'(\omega)$ corresponds to the capacitance of the cell. $C''(\omega)$ is the imaginary part of the capacitance $C(\omega)$. It corresponds to energy dissipation by an irreversible process that can lead to the movement of the molecules.

The complex power also can be calculated using the classical form:

$$S(\omega) = P(\omega) + j \times Q(\omega) \tag{3.26}$$

This equation comes directly from the definition of a complex number. P is called the active power (watt) and Q the reactive power (volt-ampere-reactive, VAR). Using the complex capacitance from Eqs. (3.24), (3.25) leads to following expressions:

$$P(\omega) = \omega \times C''(\omega) \times |\Delta V_{rms}|^2 \tag{3.27}$$

$$Q(\omega) = -\omega \times C'(\omega) \times |\Delta V_{rms}|^2 \tag{3.28}$$

where $\Delta V_{rms} = \Delta V_{rms}/\sqrt{2}$ is the maximum amplitude of the electrical signal.

The normalized imaginary part of power $|Q|/|S|$ increases when the frequency increases. When the maximum of $|Q|/|S|$ is reached, the supercapacitor behaves like a pure capacitance. However, when the power of $|P|/|S|$ is 100% at low frequency, the supercapacitor behaves like a pure resistance, that is, power is dispersed into the system and then $|P|/|S|$ decreases when frequency increases. The crossing of two plots appears when $|P| = |Q|$, referring to the time constant τ_0. The calculated time constant was found to be equal to 1.1×10^{-1} s. The time constant τ_0 represents a transition for the supercapacitor between a resistive behavior for frequency higher than $1/\tau_0$ and a capacitive behavior for frequencies lower than $1/\tau_0$. Hence, the observed time constant value of 1.1×10^{-1} s indicated that the present system can be efficiently used at low frequencies.

Galvanostatic Charge-Discharge Fig. 3.24 shows the charge-discharge profile of the supercapacitor as measured by the galvanostatic method at a constant current density of $2\,mA\,cm^{-2}$ between 0 and 1 V.

The curve profile clearly indicates typical capacitor behavior with low equivalent series resistance and IR drop. The specific capacitance for the sample was found to be 16.66 and 15.23 F g^{-1} for the first and 1000th cycle, respectively, with an efficiency of 97% and 94%, respectively. The results of the electrical parameters, namely specific capacitance (SC), specific power (SP), and specific energy (SE), are presented in Table 3.2.

(Sudhakar YN, Selvakumar M, Bhat DK. $LiClO_4$-doped plasticized chitosan and poly(ethylene glycol) blend as biodegradable polymer electrolyte for supercapacitors. Ionics (Kiel) 2013;19(2):277–85.)

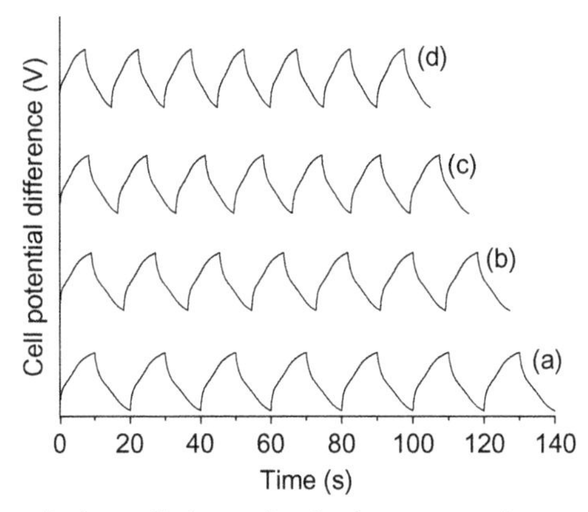

FIG. 3.24 Galvanostatic charge-discharge plots for the supercapacitor.

TABLE 3.2 Supercapacitor Parameters

Cycle No.	SC (Fg^{-1})	SP (Wg^{-1})	SE (Whg^{-1})	N (%)	ESR (Ω)	IR drop (V)
1	16.66	0.375	37.5	97	09	0.03
1000	15.23	0.372	37.2	94	12	0.04

3.4.2 Gel Biopolymer (GBPE) Electrolytes as a Supercapacitor

Major attention is being paid these days to fabricating all solid-state capacitors using solid polymer and gel polymer electrolytes because of their high ionic conductivity and advantageous mechanical properties. But solid polymer electrolytes exhibit lower ionic conductivity than gel polymer electrolytes (GPE), due to the lack of proper electrode-electrolyte contact throughout the volume of the active electrode material [48]. This distinctive advantage of better ionic conductivity is due to the fact that GPEs possess both the cohesive property of solids and the diffusive property of liquids. Although the GPEs comprise better mechanical strength for suitable applications, the phase separation that can occur between solid and liquid components within the polymer matrix still remains unsolved. GPEs employing poly(vinyl alcohol) (PVA), poly(vinyl chloride), poly(ethylene oxide) (PEO), poly(vinylidene carbonate), poly(vinylidene fluoride) (PVdF), etc., have been extensively studied [49] for improving the properties and reducing the cost to gain the attention of industries for future investment. Nonetheless, polymers that are synthesized from nonrenewable sources are facing problems of exhaustion as well as an urgent need for efficient, clean, and sustainable sources of energy. Further, there is a greater emphasis now on the use of ecofriendly materials, popularly known as green materials, everywhere. Mixing a polymer with an alkali metal salt dissolved in an organic solvent resulted in the formation of a gel-polymer electrolyte (GPE). GPEs exhibit liquid-like ionic conductivity while maintaining the dimensional stability of a solid system.

Natural polysaccharides are composed of many monosaccharide residues that are joined to each other by *O*-glycosidic linkages. When all the monosaccharides in a polysaccharide are the same type, it is called homopolysaccharide. However, when more than one type of monosaccharide is present, these are called heteropolysaccharides. Polysaccharides such as pectin play a major role in the structural integrity and mechanical strength of plant tissues by forming a hydrated, cross-linked, three-dimensional network. Cellulose is an essential ingredient of the cell wall in higher plants, and is the most abundantly available biopolymer present in nature. Another very important classification of polysaccharides is gums. Gums are present in huge quantities in varieties of plants and animals as well as marine and microbial sources. The different available polysaccharides can be classified based on their availability. We have selected three different gums for use as gel polymer electrolytes: guar gum, gellan gum, and xanthan gum.

Guar gum (GG) is a hydrophilic, nonionic polysaccharide extracted from the endospermic seed of the Cyamopsis tetragonalobus plant. GG belongs to the large family of Galactomannans. Attractive features such as low price, biodegradability, and derivation from a renewable resource make GG one of the most promising potential materials for use as a biopolymer [50]. GG hydrates in cold water to form a highly viscous solution in which the single polysaccharide chains interact with each other in a complex way [51]. GG is mainly used as an excellent stabilizer and binding agent in beverages, suspensions, textiles, explosives, papers, petroleum, and ice cream, especially in the presence of a protein-stabilizing agent. It also helps improve the quality of frozen foods [52]. The unique property of being able to sustain the gel-like character even at low temperature has encouraged us to use GG as a gel polymer electrolyte in the supercapacitor. Most commonly, fluorinated thermoplastic binders such as polytetrafluoroethylene or polyvinylidine fluoride are used in supercapacitors. These fluorinated thermoplastic binders have the disadvantage of being expensive or requiring toxic solvents. It is known that GPEs are multicomponent systems wherein the interactions between the plasticizer and polymer—as well as plasticizer and salt—influence their properties. Our preliminary studies using nonaqueous plasticizers such as propylene carbonate, ethylene carbonate, and diethyl carbonate showed noncompatibility with GG, as phase separation occurred and aggregated membranes were formed. Glycerol has been chosen as a plasticizer in this work because of its useful properties, such as practical nonvolatility at normal use temperatures and little change of the relative vapor pressure of glycerol solutions up to 70°C changes in temperatures [53]. Further, hydroxyl groups of glycerol can also facilitate the dissociation of the doped salt and maintain an ion-conducting viscous channel between the polymer matrices (Fig. 3.25).

Gellan gum (GeG) is an anionic polysaccharide composed of repetitive units of tetrasaccharides consisting of two glucoses, one glucuronic acid, and one rhamnose ring. The important properties of gellan gum gels are their high thermal stability up to 120°C, wide pH stability, and biodegradability [54]. It is believed that due to the presence of salt ions and optimum pH, the gelling and melting temperatures of gellan gels shift to higher temperatures. GeG is known to be stable in a low pH as it forms a colloidal solution. Compared with polymer electrolytes derived from the nonrenewable sources, GeG is one of the most important renewable bacterial exopolysaccharides [55]. Therefore, GeG was chosen as the host gel polymer. The stability of GeG gel was maintained by adding borax. Borax is a well-known cross-linking agent used with polymers

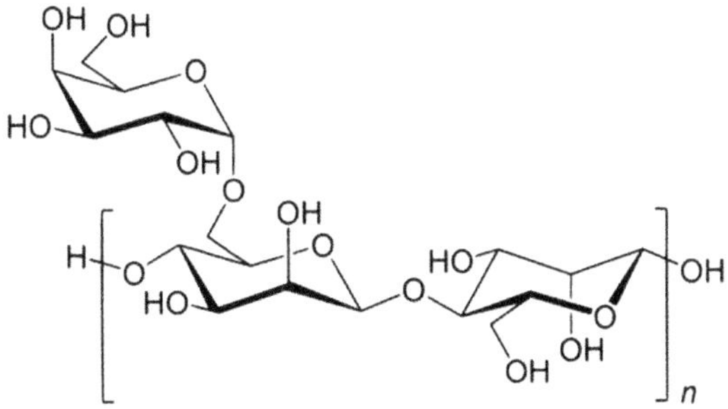

FIG. 3.25 Structure of guar gum.

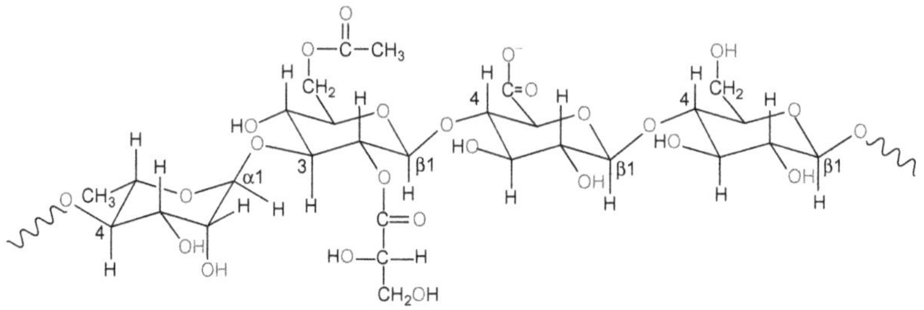

FIG. 3.26 Structure of gellan gum.

containing hydroxyl groups; it is also used as a buffer. It has been used as a cross-linker for other polysaccharides such as guar gum and scleroglucan wherein intra- and intermolecular interactions of polymer chains have been greatly increased and have good solvent retention ability under a swollen state (Fig. 3.26).

Xanthan gum (XG) is a biopolymer with branched chains. It is obtained from the microbiological fermentation in aerobic conditions of sugar cane, corn, or their derivatives, which are transformed into a soluble gum during the reaction in the presence of the bacterium Xanthomonascampestris. The resultant gum turns into xanthan gum powder by precipitation in a nonsoluble solvent. XG consists of D-glucosyl, D-mannosyl, and D-glucuronyl acid residues in a 2:2:1 M ratio and variable proportions of *O*-acetyl and pyruvyl residues. It is an acidic polymer consisting of pentasaccharide subunits, forming a cellulose backbone with trisaccharide side chains composed of mannose- (β-1,4)- glucuronic acid- (β-1,2)-mannose attached to alternate glucose residues in the backbone by α-1,3 linkages [56]. XG in its solution state provides uniform viscosities over the temperature range of freezing to near boiling with excellent thermal stability. XG has excellent solubility and stability under both acidic and alkaline conditions, even with salts. Due to its extraordinary properties, XG is used in the food, cosmetics, and pharmaceuticals industries (Fig. 3.27).

A Case Study on Lithium Salt-Doped Guar Gum as a Gel Polymer Electrolyte for a Supercapacitor

Experimental

Material Preparation Guar gum (GG) (medium molecular weight) and glycerol were purchased from Merck. Lithium perchlorate (Aldrich) was dried at 393 K and kept under vacuum for 48 h before use. A stock solution was prepared by dissolving 2 g of GG in 100 mL of distilled water. The GPEs were prepared by mixing the appropriate amount of salt and plasticizer and labeled as S1, S2, S3, S4. These samples contain 90 wt.% of GG stock solution; 10 wt.% of glycerol (optimized as it showed better plasticizer retention property at this concentration) and varying $LiClO_4$ as

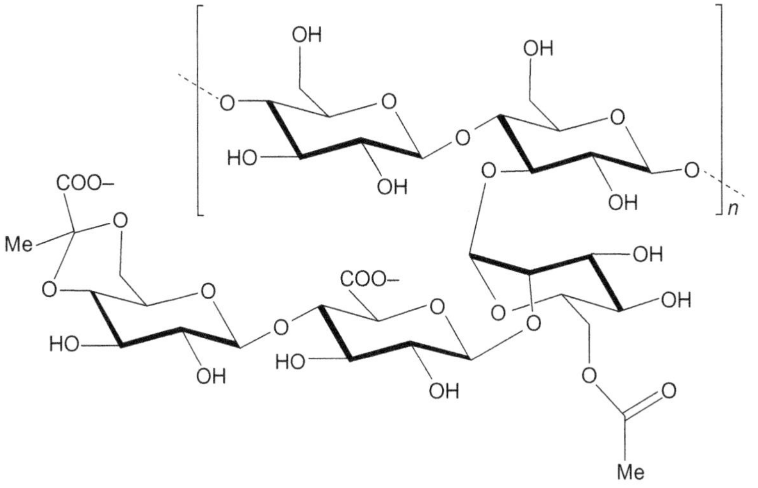

FIG. 3.27 Structure of Xanthan gum.

0.075 wt.%, 0.10 wt.%, 0.25 wt.% and 0.50 wt.%, respectively. The so-prepared four solutions were placed in clean 10 mL beakers separately and allowed to bring about gelation initially at room temperature. They were then kept in a water bath at 333 K for 48 h to form GPEs before being subjecting to other studies.

Characterization Fourier-transform infrared-spectroscopic (FTIR) measurements of the undoped and $LiClO_4$-doped GG samples were carried out at room temperature using a Nicolet Avatar 5700 FTIR spectrometer. Differential scanning calorimetery (DSC) and thermal gravimetric analysis (TGA) measurements of the undoped and $LiClO_4$-doped GG samples were done on DSC-60 and DTA-60 model instruments from Shimadzu, respectively. Measurements were performed over a temperature range of 303–473 K at a heating rate of 10°C min^{-1} under the nitrogen atmosphere at a flow rate of 60 mg L^{-1}. Readings were taken from the first heating run.

The solid GPE sample was cut into small cubes and freeze dried in a deep freezer. Later, freeze-dried samples were kept in a nitrogen atmosphere and then subjected to high vacuum. The microimages were taken using a scanning electron microscope (SEM), the ZEISS EVO 18 special edition.

For electrochemical studies, samples having a thickness of ~2 mm were cut into 1 cm × 1 cm square dimensions and placed between two square copper electrodes (length 1 cm) fitted with copper wires. The whole setup was held tightly with a plastic clamp. The bulk ionic conductivities (σ) and dielectric properties of the blends were determined from the electrochemical impedance spectra (EIS) in the frequency range between 1 to 100 MHz using a small amplitude AC signal of 10 mV. Experiments were carried out in the temperature range of 303–333 K using a PID-controlled oven from SES instruments Pvt. Ltd. From these, AC impedance data dielectric studies were carried out.

Fabrication of a Symmetrical Supercapacitor Cell The electrode material for supercapacitor fabrication was prepared using activated carbon (AC) derived from areca fibers that have a surface area of $250\,m^2\,g^{-1}$. The AC was coated on two stainless steel electrodes using GG as a binder (GG/AC in 0.2:1 weight ratio). The supercapacitor cell was constructed using GPE sandwiched between two prepared AC-coated electrodes. The unit cell was sealed in a plastic coated aluminum pouch, keeping the two wires outside. Electrochemical characterization was carried out by CV, EIS, and GCD studies. For lower temperature studies (273–293 K), the sealed unit cell was placed in a water-circulated, low-temperature bath, keeping the wires above water and clamped tightly while allowing it to wait for a few minutes to attain the applied temperature (temperature error of ±1°C). AC impedance readings were then taken. All the electrochemical studies were carried out using a BioLogic SP-150 instrument.

Results and Discussion

FTIR studies The FTIR spectra (Fig. 3.28) of pure GG show a broad peak at $3328\,cm^{-1}$ due to O—H stretching and, on the addition of lithium salt, it shifted to $3287\,cm^{-1}$. The other important peaks at 1034, 1092, and $1160\,cm^{-1}$ corresponding to C—O—C stretching from glycosidic linkages and O—H bending [57] are broadened, indicating the presence of functional group interactions upon the addition of glycerol. The characteristic peak of $LiClO_4$ at $628\,cm^{-1}$ [21] disappeared and new peak is observed for the doped sample at $662\,cm^{-1}$. The shifting of peak and the appearance of a new peak in the polymer electrolyte system suggest the existence of interaction between GG, salt, and the plasticizer.

DSC and TGA Studies Fig. 3.29 shows the DSC thermograms of GPE samples containing glycerol along with $LiClO_4$ (S1–S4) and pure GG. In the thermogram of pure GG, the bounded water exhaustion and melting peak are observed at 136°C and 225°C, respectively. On addition of glycerol and $LiClO_4$, a well-distinguishable shift in peaks and enthalpy changes is observed. As the concentration of lithium salt was increased, the water retention ability of samples (S1–S4) decreased from

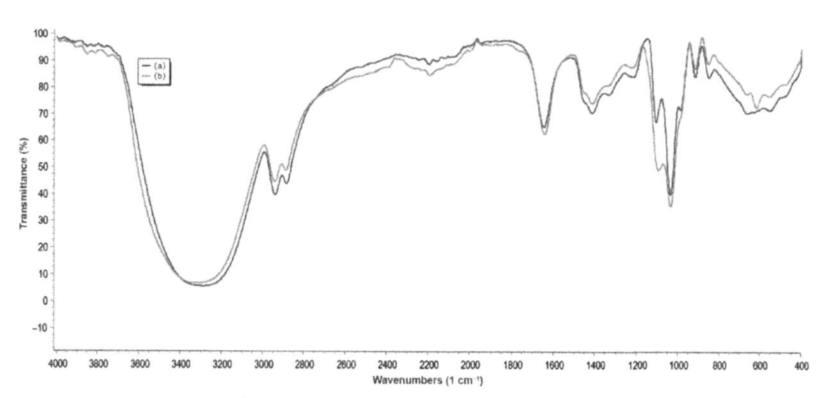

FIG. 3.28 FTIR spectrum of (a) pure GG and (b) S2.

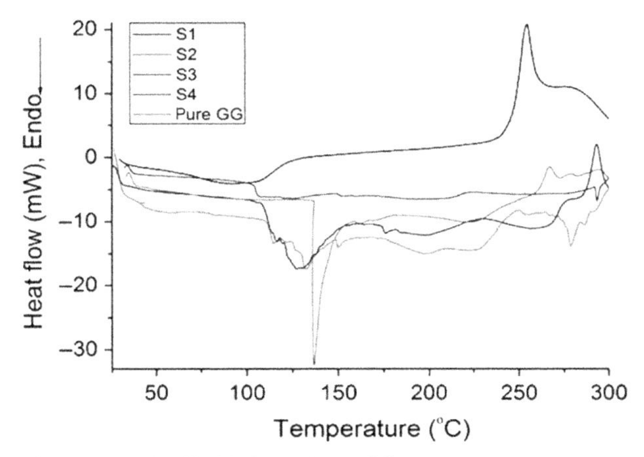

FIG. 3.29 DSC results of S1, S2, S3, S4, and pure GG.

132°C to 104°C while the shift in the melting peak was not significant (197–196°C). This may be due to the low pressure glycerol. This indicates that glycerol is compatible with GG and the added salt has formed bonds with GG.

The amount of heat absorbed during transitions in polymer signifies the stability of the system at varying temperatures. The heat absorbed during the first transitions is reduced from −199 mcal (pure GG) to −60 mcal (S1) while samples S2, S3, and S4 absorbed −13, −62, and −70 mcal, respectively. Hence, the samples containing glycerol and $LiClO_4$ were proved to be stable and could be used as a gel electrolyte up to 278°C. A further increase in temperature results in oxidation and decomposition of the GPE.

To further elucidate the results of DSC, TGA was performed for the pure GG and S1–S4 samples (Fig. 3.30). In pure GG, the initial weight loss that began at 94°C is due to the presence of a small amount of moisture in the sample. This continues due to the slow release of solvent water from the bulk of the polymer. The second degradation region is at 260°C where the polymer decomposition starts and continues up to 326°C. For the plasticized $LiClO_4$-doped system, the onset of degradation temperature decreased with increasing salt content (S1–S4). For the samples S1, S2, S3, and S4, after the initial loss of weight due to the presence of some moisture, there is continuous weight loss that starts from 222°C, 220°C, 201°C, and 197°C, respectively. Thus, it can be concluded from TGA results that plasticizing and doping of salt have significantly affected the thermal stability of the GPE system to some extent in comparison with pristine GG. This also supports the conclusion derived from the FTIR data that there is some interaction between polymer and doped salt.

Differential thermal analysis (DTA) was performed for the pure GG and S1–S4 samples (Fig. 3.31). In pure GG, the initial weight loss that began at 94°C is due to the presence of a small amount of moisture in the sample. This continues due to the slow release of solvent water from the bulk of the polymer. The second degradation has a broad peak region at 260°C where the polymer decomposition starts and continues up to 326°C. For the plasticized $LiClO_4$-doped system, the onset of degradation temperature decreased with increasing salt content (S1–S4). For the

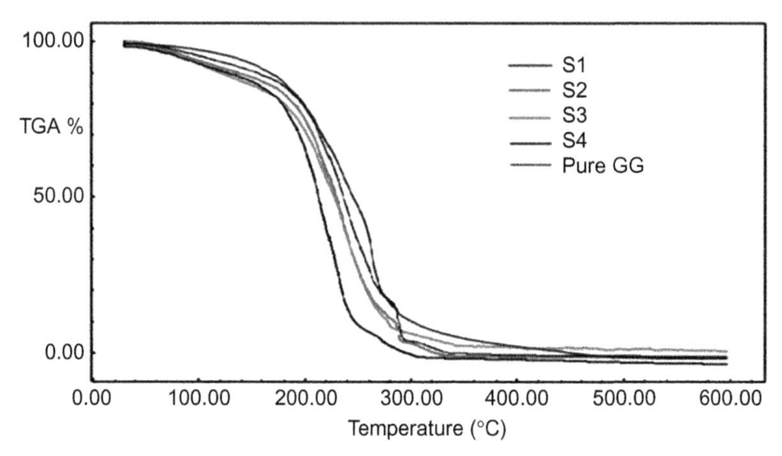

FIG. 3.30 Thermal gravimetric analysis results of S1, S2, S3, S4, and pure GG.

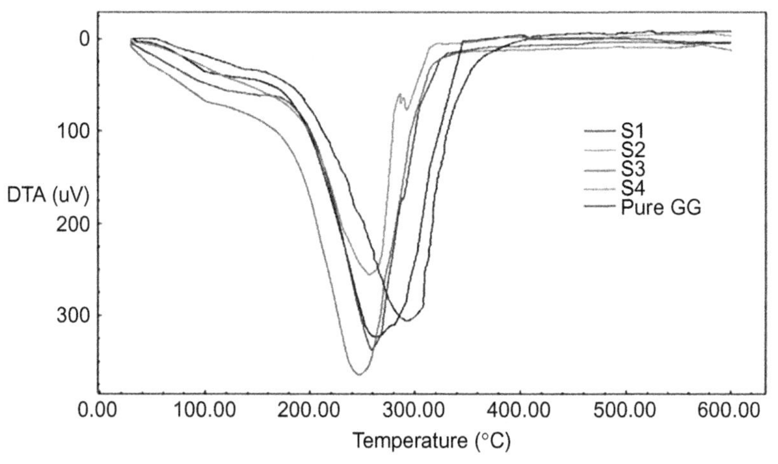

FIG. 3.31 Differential thermal analysis (DTA) results of S1, S2, S3, S4, and pure GG.

samples S1, S2, S3, and S4, after an initial loss of weight due to the presence of some moisture, there is continuous weight loss that starts from 222°C, 220°C, 201°C, and 197°C, respectively. Thus, it can be concluded from DTA results that plasticizing and doping of salt have significantly affected the thermal stability of the GPE system to some extent in comparison with pristine GG.

SEM Studies Fig. 3.32A and B show the SEM images of the GPE sample S2. As can be seen from the figures, the gel polymer exhibits unusual tubular array-like morphology unlike the common gel polymers, which exhibit microporous-, fractured-, globular-, or planar folding-type features. To our knowledge, this new observation of the tubular arrangement of the gel polymer matrix is unreported and may open a new direction for further research. The tubular structures had a dimension of

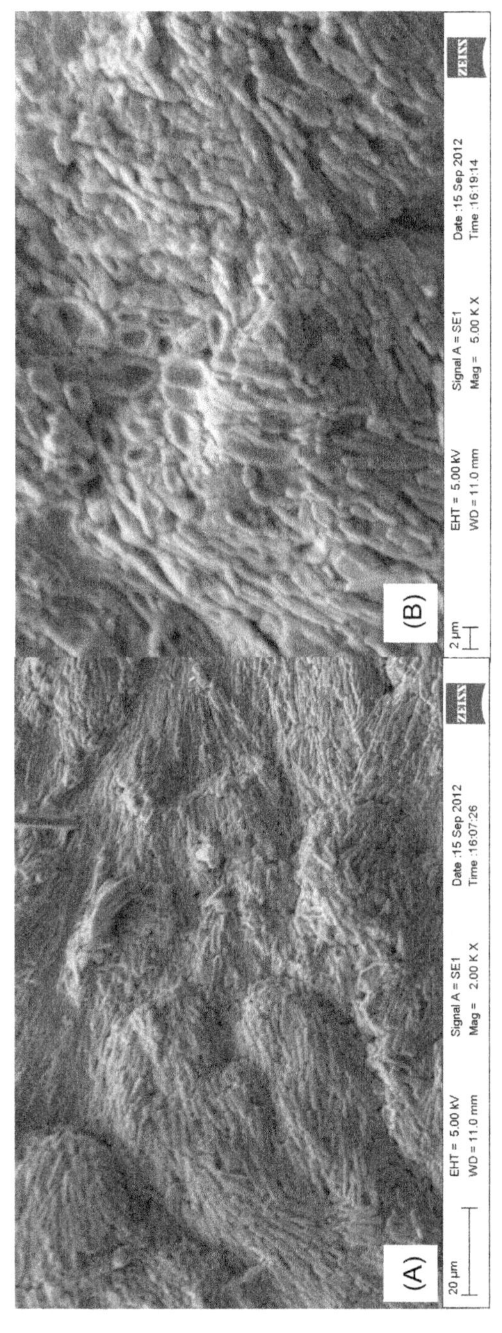

FIG. 3.32 (A and B) SEM images of the S2 gel polymer electrolyte at different magnifications.

~5–15 μm in length and 1–4 μm in diameter. The pore sizes were not larger than 5 μm, indicating that the gel polymer has been formed by slow drying [58]. Structures with small pores are known to have higher liquid electrolyte retention ability. Further, the tubular arrays may provide an ideal channel for conduction of Li^+ ions. These tubules in viscous media are free to move as the temperature increases by self-adjusting or by breaking/forming hydrogen bonds in the gel system. This might be the reason why, in the TGA studies, the effect of temperature was not big when doped with salt and plasticizer. Interestingly, a recent SEM study on arrays of $Pt@RuO_2$ core-shell nanotubes [59] having 10–20 μm in length was also similar to the surface morphology of GPE presented here. Hence the same type of conduction mechanism in the GPE can also be expected. Further, the spherulite form of the surface was not observed, indicating that lithium salt has been dissociated within the polymer matrix and holds a good interaction between plasticizer and polymer.

Ionic Conductivity Studies The ionic conductivity decreases with an increase of the dopant $LiClO_4$ (Fig. 3.33). The regression value, R^2, for the plot of log σ versus $1000/T$ ranged from 0.95 to 0.98, which is almost unity, suggesting that the plot can be considered linear. This obeys the Arrhenius equation, where the ionic transport is promoted by the viscous medium. E_A was calculated from the slope of the log σ versus $10^3/T$ graph (Fig. 3.33) and depicted in Fig. 3.34.

The ionic conductivity has been found to be $3.02 \times 10^{-2}\,S\,cm^{-1}$ at 333 K, and $2.2 \times 10^{-3}\,S\,cm^{-1}$ at 303 K for sample S2. As the temperature increases, the polymer chains may acquire faster internal modes of vibration in which bond rotations produce segmental motion, favoring interchain and intrachain lithium ion-hopping movements on the oxygen atom of GG and hence the increase in the conductivity of the polymer electrolyte. It may be noticed that the increase in lithium ionic conductivity at a higher GG content is related to better initial plasticizer retention in tubular structures. S1 showed lower ionic conductivities because the low concentration of $LiClO_4$ could not break the intermolecular bonding between GG chains. Later, a decrease in ionic conductivity was observed with an increase in salt concentration because the ionic motion was hindered within the pores of the tubular structure and only the plasticizer retained between the tubular polymer matrix

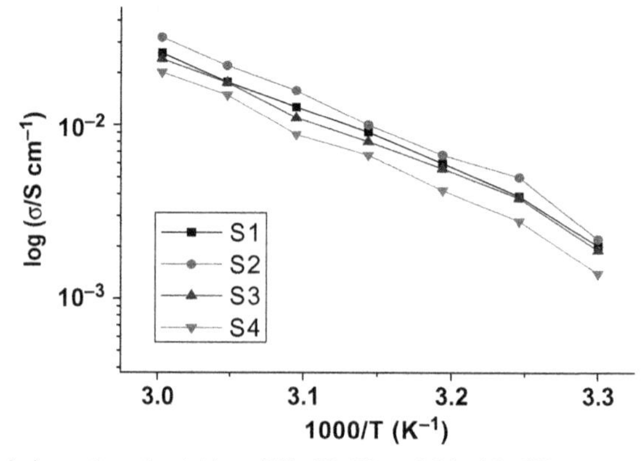

FIG. 3.33 Variations of conductivities of S1, S2, S3, and S4 with different temperatures.

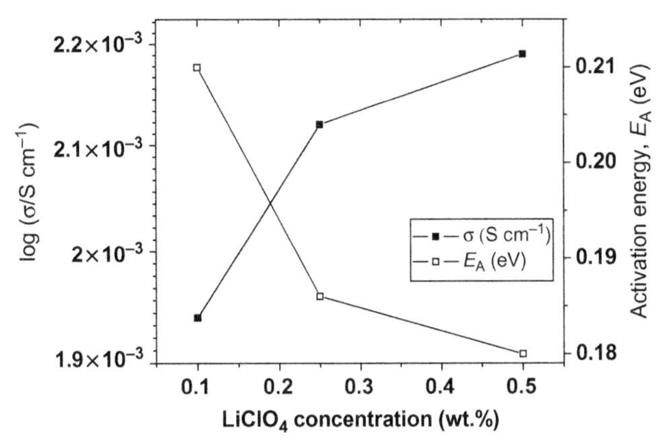

FIG. 3.34 Variation of activation energy (E_A) and room temperature conductivity as a function of $LiClO_4$ concentration.

helps in the conduction of ions. Adding more and more cations may lead to the formation of higher ionic aggregates, leaving a lower number of free cations.

Activation energy is the energy required for an ion to initiate movement. When the ion has acquired sufficient energy, it is able to hop from the donor site and move to another donor site. It can be observed from Fig. 3.34 that the values of activation energy were opposite to the conductivity. The values of E_A for GPEs studied in this work are in the range of 0.21–0.18 eV. The electrolyte with the lower value of E_A implies that dopant salt has been dissociated, favoring ionic conduction by forming coordination with other polymer sites. The decrease in E_A with an increase in salt indicates the formation of ion pairs or aggregates, hence decreasing the conductivity. Comparatively, the present ionic conductivity is better than other ionic conductivities exhibited by natural polymer electrolytes, which ranged from 10^{-3} to 10^{-6} S cm^{-1}. Furthermore, conventional gel electrolytes based on acid-doped poly(methylmethacrylate) in dimethyl-formamide showed conductivity around 10^{-5} S cm^{-1} at room temperature, which is lower than gel containing a large amount of water.

As shown in Fig. 3.35, the bulk resistance of the GPE at higher frequencies decreases with an increase in temperature. At a low-frequency region, initially at 303 K, there is vertical response, which is characteristic of ideal response and might be due to the presence of a large number of ions near the electrode/electrolyte interface containing the plasticizer-rich region. As the temperature increases, there is a decrease in slope, demonstrating the dominance of the Warburg impedance, which is characterized by a vertical line at the phase angle 45 degrees, wherein the diffusion of ions into the bulk of the electrodes takes place. At 333 K, the bulk resistance decreased drastically, leading to random orientations of dipoles in the side chains and even rupturing the weak hydrogen bonding between glycerol and GG. Further, the reduction in the vertical line of the Warburg impedance suggests that there is an increase in the liquid component in the gel system.

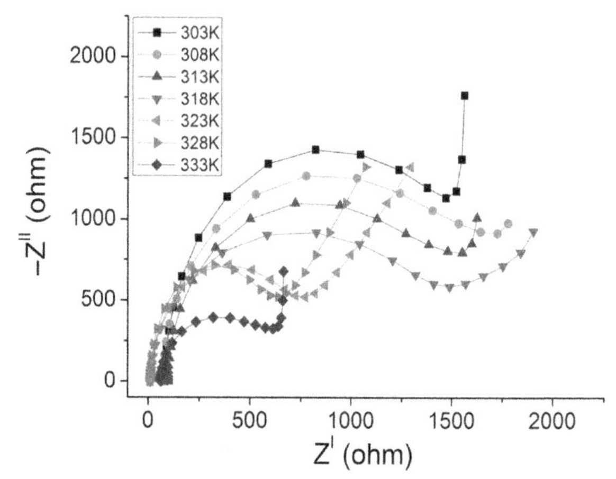

FIG. 3.35 Nyquist impedance plots of the sample (S2) biodegradable GPE at different temperatures.

Dielectric Studies A broad frequency range of dielectric relaxation spectroscopy is a very useful tool to study the relaxation of dipoles in the plasticized polymer electrolytes. It also yields information on the movement of charges through that gel. The SEM images that were taken for the freeze-dried and vacuumed samples revealed tubular structures, impling that these tubular structures in the gel state were more relaxed and free to move in the plasticizer medium along with the doped lithium ion. Because water was used as the solvent, it will have an effect on GP. Therefore to clarify these arguments, dielectric studies were done. Due to the fast movement of ions in the GPE, the ions possess less time for dielectric relaxation. This response in the blocking electrode is less assessable compared to the copper electrode wherein the copper electrode does not perfectly block the lithium ions and also possesses better surface roughness. The variation of real and imaginary parts of the dielectric constant as a function of frequency at 303 K for GPE sample S2 is shown in Figs. 3.36 and 3.37, respectively.

The trend is equally similar in both the figures, where it rises sharply toward the low frequency and decays at an increasing frequency. This is attributed to the electrode polarization effect wherein charge accumulation decreases with increasing frequency, thereby decreasing both the dielectric constant (ε_R, real part) and the dielectric loss (ε_I, imaginary part). As frequency increases, the ε_R decreases due to the high periodic reversal of the electric field at the interface, which reduces the contribution of charge carriers toward the dielectric constant. Furthermore, the dispersion shifts toward the high-frequency region as the temperature increases. Therefore, from these studies it can be inferred that in the present GPE, the plasticizer has increased the ionic mobility by reducing the potential barrier to ionic motion by decreasing the cation-anion coordination of salt. Also, the absence of conductivity relaxation peaks at the high-frequency region indicated that local movement of dipole groups was absent. The number of free Li^+ ions in the tubular array was less; therefore, the intensity of the peak observed at the

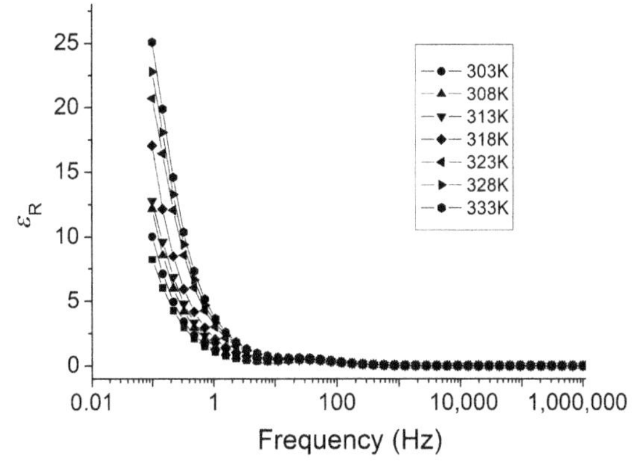

FIG. 3.36 Plots of dielectric constant versus frequency at different temperatures.

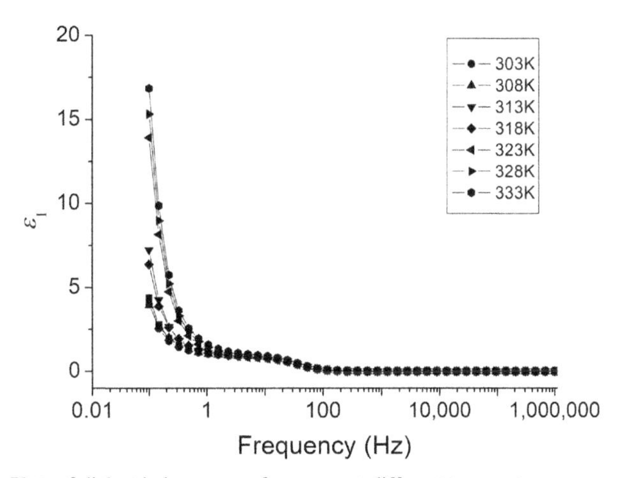

FIG. 3.37 Plots of dielectric loss versus frequency at different temperatures.

middle-frequency region was smaller. As the temperature increased, the peaks increased, indicating more ions were pulled from these tubular structures. Thus, the increase in mobile ions increased the conductivity in view of the fact that the conductivity is proportional to the number of mobile ions.

The advantages of modulus studies are to recognize the conduction process and the effect of the frequency. The contribution of electrode polarization phenomena can be ignored at the lower frequency of the real part of the electric modulus (M_R) and the imaginary part (M_I), which are represented in Figs. 3.38 and 3.39, respectively.

The peaks in the modulus formalism at high frequencies show that the present GPE is an ionic conductor. A tail was observed in the lower part of the frequency, which indicates that the sample is capacitive in nature. The presence of the

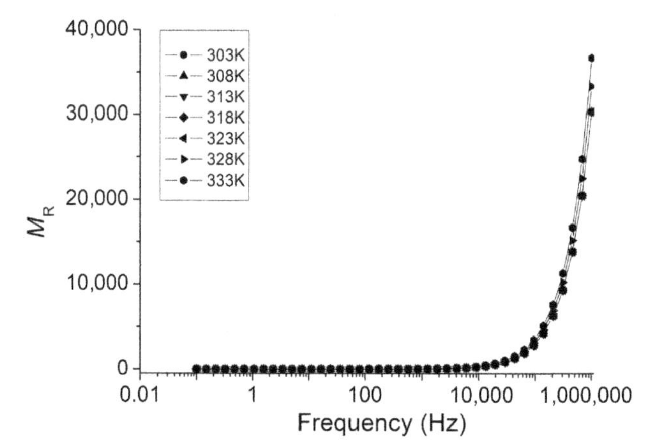

FIG. 3.38 Plots of the real part of electric modulus versus frequency at different temperatures.

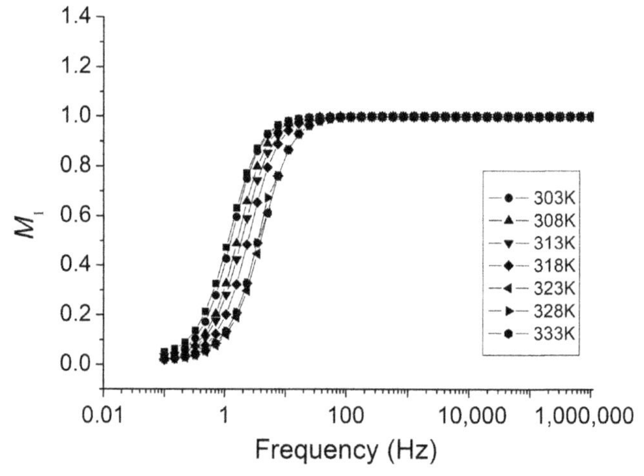

FIG. 3.39 Plots of the imaginary part of electric modulus versus frequency at different temperatures.

relaxation process in M_I indicates residual water contributes toward conductivity enhancement. Consequently, it may be suggested that the charges causing these effects are closely associated with the gel itself, rather than existing in a free state within the network. This is compatible with the conclusions drawn from FTIR and thermal studies. The peak curve at higher frequencies may be due to the bulk effect, whose mechanism will be discussed later.

Supercapacitor Studies GPE with higher conductivity (S2) was used to construct a supercapacitor. Cyclic voltammetry (CV) responses for the carbon-carbon symmetrical supercapacitor at various sweep rates are shown in Fig. 3.40.

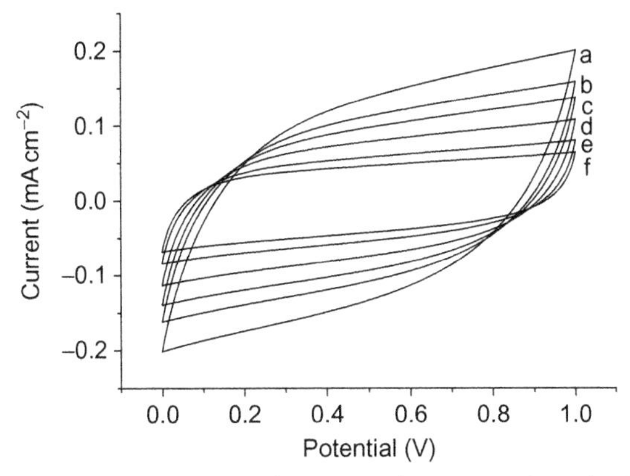

FIG. 3.40 CVs of the fabricated carbon-carbon symmetric supercapacitor using (S2) GPE at scan rates. (a) $50\,mV\,s^{-1}$, (b) $25\,mV\,s^{-1}$, (c) $20\,mV\,s^{-1}$, (d) $15\,mV\,s^{-1}$, (e) $10\,mV\,s^{-1}$, and (f) $5\,mV\,s^{-1}$.

The specific capacitance values of the supercapacitor have been calculated from the respective cyclic voltammograms. A maximum specific capacitance of $186\,F\,g^{-1}$ at a scan rate of $5\,mV\,s^{-1}$ was obtained, which is more than the capacitance of hydrogel systems reported by Choudhury et al. [60]. This implies that large amounts of mobile Li^+ are available in the gel, which leads to a high capacitance of the device. The CV curves have an almost rectangular shape and a nearly mirror-image symmetry of the current response about the zero current line. These features indicate the capacitive behavior of the supercapacitor, that is, the double-layer formation at the interfaces.

To investigate the change of the resistance of components, electrochemical impedance spectroscopy (EIS) at different temperatures was measured. Fig. 3.41 shows the AC impedance response of the symmetrical supercapacitor fabricated using activated carbon and GPE.

In Fig. 3.41A, the EIS spectra for activated carbon show good performance. When it was fabricated as a supercapacitor using GPE, small oscillations were observed in EIS at the lower-frequency region and the R_{ct} increased from 5 to $18\,\Omega$. This might be due to the tubular array in the biodegradable GPE. In the range between 273 and 293 K, the spectra remained identical in nature. As expected, only the high-frequency resistance shifts to higher value as the temperature decreased. The semicircle results from the parallel combination of resistance and capacitance and the linear region is because of the Warburg impedance. Charge transfer resistance (R_{ct}) is generally due to contact resistance between the electrode material and the current collector and interparticle resistance of AC. As observed, the R_{ct} values did not increase significantly with a decrease in temperature. The R_{ct} values were 14, 17, 18, and $18\,\Omega$ at 293, 283, 278, and 273 K, respectively. The C_{dl} values obtained were 16, 17, 17.6, and $17.7\,\mu F\,cm^{-2}$ at 293, 283, 278, and 273 K, respectively. These results point toward the usage of GPE for application in supercapacitors, even at lower temperatures. The property of glycerol that maintains a viscous nature even at low temperature helps the tubular array of the polymer to be stable as temperature decreases. At higher temperatures,

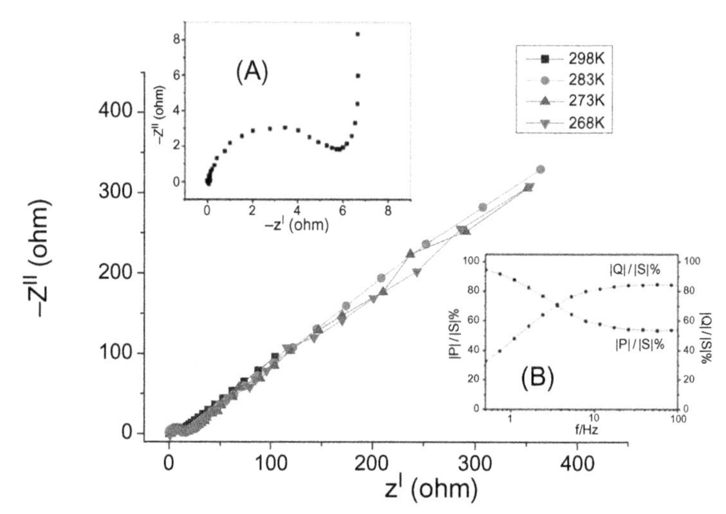

FIG. 3.41 AC impedance plot of the supercapacitor at various temperatures (inset: (A) AC Impedance plot of activated carbon, (B) Plot of normalized reactive power |Q|/|S|% and active power |P|/|S|% versus frequency (Hz) of the supercapacitor at 293 K).

the accessibility of ionic motions is increased in both AC-coated electrodes and GPE, leading to a further decrease in the semicircle at the higher-frequency region and an increase in the straight line at the lower-frequency region. Hence, the interaction of electrode/electrolyte interface increases, which results in the good electrochemical property.

Using a normalized imaginary part (reactive power) $|Q|/|S|$ and a real part (active power) $|P|/|S|$ of the complex power versus frequency plot for the supercapacitor at 293 K has been calculated and is shown in Fig. 3.41 (inset). The calculated time constant was found to be equal to 2 s.

Galvanostatic charge-discharge measurements were performed to study the influence of the ionic conductivity on the performance of the constant current charge and discharge characteristics of the SCs. Fig. 3.42 shows the charge-discharge profile of the supercapacitor as measured by the galvanostatic method by varying current densities of 2, 4, 6, 8, and 10 mA cm^{-2} between the potential window of 1 V with respective decreasing discharge capacitance (C_d).

All curve profiles clearly indicate typical capacitor behavior. Specific energy (SE) and specific power (SP) values ranges from 15 Wh kg^{-1}and 1950 W kg^{-1} to 25 Wh kg^{-1} and 2680 W kg^{-1}, respectively, as observed in the Ragone plot (Fig. 3.42, inset). Furthermore, the galvanostatic charge-discharge stability of GPE was measured at a constant current density of 2 mA cm^{-2} between the potential widow of 1 V, that is, between 0 and 1.0 V up to 2000 cycles (Fig. 3.43).

The C_d for the sample was found to be 164 and 155 mF cm^{-2} for the first and 1000th cycle with an efficiency of 98% and 96%, respectively. Equivalent series resistance (ESR) was 51 and 67 Ω for the first and 1000th cycle, respectively, with IR drop of 0.02 and 0.03 V, respectively. The SP and SE were 2600 W kg^{-1} and

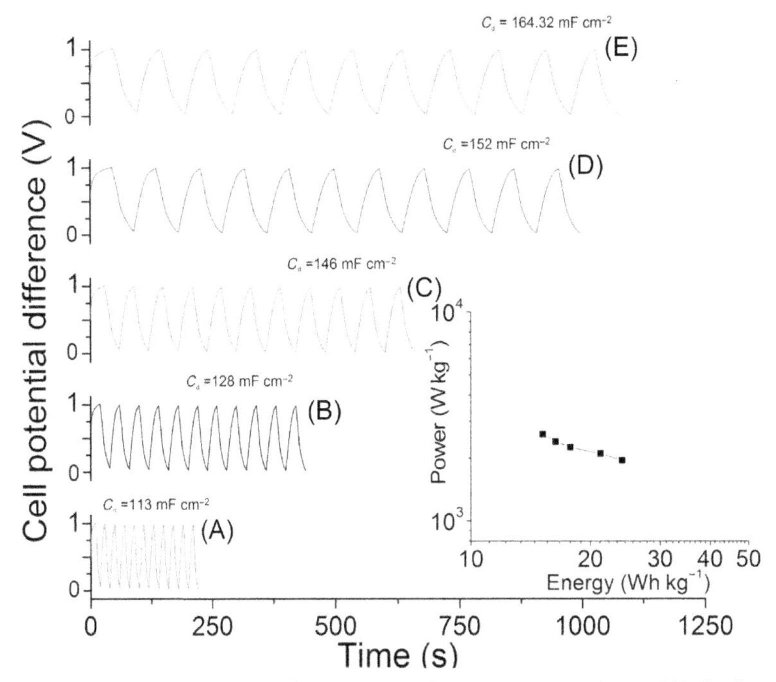

FIG. 3.42 Galvanostatic charge-discharge plots for the supercapactior at (A) $2 mA cm^{-2}$, (B) $4 mA cm^{-2}$, (C) $6 mA cm^{-2}$, (D) $8 mA cm^{-2}$, and (E) $10 mA cm^{-2}$ (inset: Ragone plot related to power and energy densities of supercapacitor).

$24 Wh kg^{-1}$, respectively. The initial decrease in the capacitance values is minimum, which is due to the loss of the charges initially stored at interfaces associated with the irreversible reactions of loosely bound surface groups on the porous activated carbon electrodes. Also, the advantage is that GG itself is used as a binding agent for activated carbon, which certainly improves the ionic diffusion through the pores of activated carbon until ions stabilize, hence enhancing the double-layer capacitance at the electrode/electrolyte interface. Remarkably, a supercapacitor developed using GPE offers energy density and power density performances much better than the recently reported pitch carbon-coated $Li_4Ti_5O_{12}$, nonaqueous hybrid supercapacitor ($16 Wh kg^{-1}$ and $1010 W kg^{-1}$) [61]. Further, the observed performance is better than that achieved by conventional systems such as hybrid supercapacitors using polymer materials like poly-fluoro phenyl thiophene (PFPT)/LTO ($20 Wh kg^{-1}$) [62] or PVA-KOH-KI ($7.8 Wh kg^{-1}$ and $1534 W kg^{-1}$) [63].

Based on the evidence obtained from characterization studies, the most possible mechanism of charge transport in the fabricated supercapacitor through this GPE and AC is explained in Fig. 3.44.

As observed in FTIR and thermal studies, there were interactions between GG, glycerol and salt, indicating complete dissociation of salt when the concentration is

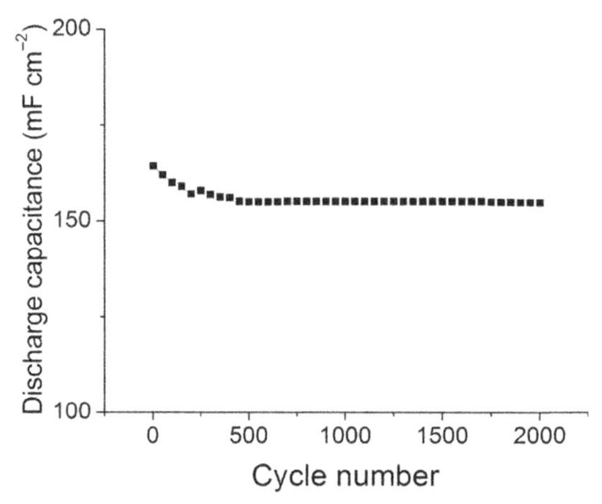

FIG. 3.43 Discharge capacitance versus number of cycles of GPEs supercapacitor (current density: $2\,mA\,cm^{-2}$).

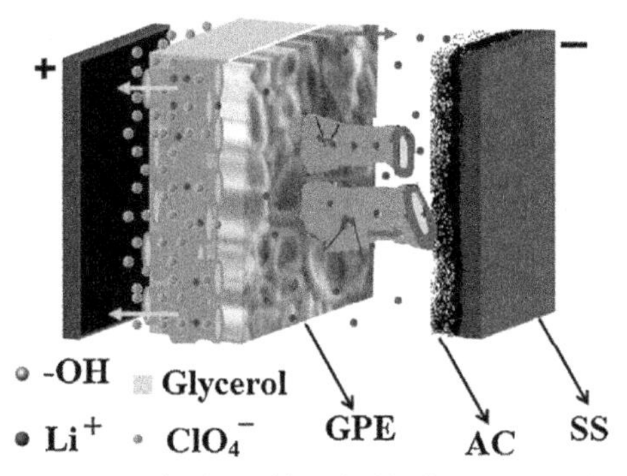

FIG. 3.44 Possible transport mechanism of ions inside the supercapacitor during the charging condition (SS, stainless steel; AC, activated carbon).

optimized in the gel system. A tubular accumulation of the gel matrix in SEM studies showed that two pathways for lithium ion conduction exist: one through these tubular structures and the other through the region surrounding them. The peaks corresponding to Li^+ ions inside these tubular structures during dielectric studies suggest that ions between tubular arrays move faster than the withheld ions inside them, which also might be due to the weak coordination between the OH groups of GG and Li^+ ions, as shown in the figure. This might be the reason for the enhanced energy density and better power density of the gel system. In addition, AC was bonded to a stainless steel current collector using GG as the binder; this material now acts as a sponge absorbing the little plasticizer component from the GPE along with salt ions. During charging, more ions are drawn inside electrode materials, which increases the double layer, hence the capacitance. During discharging, the concentration difference between the GPE and electrode material pulls back a few ions toward the gel system, leaving behind the remaining ions within the tubular structures. Once the charge-discharge cycle increases, both the GPE and electrode material stabilize with the constant current supply between fixed potential ranges. To utilize the new finding of the tubular array in GPE, a further study is much needed wherein these tubular array could be modified or doped with graphene oxide/metal oxide to produce hybrid stable GPEs. Moreover, this gel system could be doped with acids to examine their stability for application in supercapacitors.

(Sudhakar YN, Selvakumar M, Bhat KD. Tubular array, dielectric, conductivity, and electrochemical properties of biodegradable gel polymer electrolyte. Mater Sci Eng B 2014;180:12–19. ISSN 0921-5107.)

3.4.3 Solid Biopolymer (SBPE) Electrolytes as a Supercapacitor

Green materials have emerged as a challenging concept in research to satisfy sustainable development and provide future generations with the ability to avail their own needs. This motivation among researchers has led to many engineered materials in the medical field and devices such as batteries or supercapacitors in the energy field. Only a few studies are available in which either the devices as a whole or the materials in them are biodegradable [64,65]. The main materials—such as the electrode material and the electrolyte, especially the solid polymer electrolyte—play an important role in preventing leaks, providing better energy density, and having less thickness and less weight as well as being flexible compared to the liquid counterpart [66]. Furthermore, the use of liquid electrolytes needs a separator to prevent electrode-electrode contact, which drastically increases the cost.

The commonly used amorphous polymers include polyacrylonitrile, polyvinylidene difluoride, and polymethylmethacrylate [49,67], which exhibit high conductivity due to gel formation wherein the polymer network entraps the liquid electrolyte. Although desirable properties can be achieved using these petroleum-derived polymers, they are harmful to the environment in larger quantities when discarded and mainly depend on nonrenewable sources. Despite several efforts from the scientific community, there is still a lack of highly abundant and low-cost solid polymer electrolytes that avoid the use of toxic chemicals while degrading at the end of its life period. In comparison to electrolytes with alkali metal salts, proton conductors are characterized by higher dynamics of ionic transport. Dielectric analysis is an informative technique used to determine the molecular motions and structural relaxations present in polymeric materials possessing permanent dipole moments [66]. It also reveals the cause of specific capacitance in the supercapacitors.

Solid polymer electrolytes (SPEs) have been proved to be prospective candidates for advanced electrochemical device applications because of their characteristics such as viscoelasticity and flexibility as well as high ionic conductivity. Interest in these materials grew mainly because of the pioneering measurements of the ionic conductivity of polymer salt complexes reported by Wright et al. [68]. Aside from poly(ethylene oxide), a few types of biopolymers such as cellulose with its derivatives and other biodegradable polymers [69] are also used as polymer matrixes in developing various SPEs.

Arof et al. [70] have reported that films prepared from high molecular weight chitosan exhibit the highest electrical conductivity of 2.14×10^{-7} $\mathrm{S\,cm^{-1}}$. The electrical conductivity was further enhanced to $1.03 \times 10^{-5}\,\mathrm{S\,cm^{-1}}$ when ethylene carbonate (EC) was used as a plasticizer. Osman et al. [40] have investigated the cast films of chitosan acetate, plasticized chitosan acetate, chitosan acetate containing salt, and plasticized chitosan acetate-salt complexes in order to obtain some insight on the mechanism of ionic conductivity in chitosan-based polymer electrolytes. Osman et al. [71] have reported on chitosan-salt interactions. When lithium triflate was added to chitosan to form a film of chitosan acetate-salt complex, the bands assigned to chitosan in the complex and the spectrum as a whole shifted to lower wave numbers.

Chitosan with adipic acid showed a conductivity of $1.4 \times 10^{-9}\,\mathrm{S\,cm^{-1}}$, which reported that conductivity enhancement is caused not only by the increase in the concentration of free ions but also by the increase in the mobility and diffusion coefficient of ions [72]. Chitosan-Co_3O_4 composite film showed $1.94 \times 10^{-2}\,\mathrm{S\,cm^{-1}}$ along with high permittivity and dielectric at low frequency. The dielectric measurements suggested a conduction mechanism wherein the conduction is mainly based on the electronic exchange between Co^{2+} and Co^{3+} ions while chitosan is an important medium in terms of stability and conduction [73]. A good amount of effort has been harnessed to produce chitosan-based polymer electrolytes [74,75].

A solid-state PSSA was demonstrated for the first time by Wee et al. [76] as an effective ion-conducting electrolyte medium in supercapacitors with electrodes based on carbon nanotube (CNT) networks, where ionic conductivity reached up to $4 \times 10^{-4}\,\mathrm{S\,cm^{-1}}$ at 80% relative humidity (RH). These polyelectrolytes tend to be rather rigid and are poor proton conductors unless water is absorbed.

Mohamed et al. [77] have reported that potassium hydroxide (KOH) disrupts the crystalline nature of PVA-based polymer electrolytes and converts them into an amorphous phase. The PVA-KOH alkaline solid polymer electrolyte system with PVA/KOH wt.% ratio of 60:40 exhibits the highest room temperature ionic conductivity of $8.5 \times 10^{-4}\,\mathrm{S\,cm^{-1}}$.

Fonseca et al. [78] have reported a new solid polymer electrolyte composed of a polymeric matrix of biodegradable polymer, poly-ε-caprolactone (PCL), and 10% $LiClO_4$ used in a rechargeable lithium polymer battery. The maximum conductivity was $1.2 \times 10^{-6}\,S\,cm^{-1}$ for 10 wt.% $LiClO_4$. A $LiNiCoO_2$ film was prepared by the combination of sol-gel synthesis and a template concept was used as the cathode. Cyclic voltammetry, charge-discharge cycle, and electrochemical impedance spectroscopy were used to characterize the Li/PCL, 10% $LiClO_4/LiNiCoO_2$ device.

Rajendran et al. [79] have studied the solid polymer electrolytes of high ionic conductivity prepared using poly(vinyl alcohol) (PVA) complexes with various lithium salts such as $LiClO_4$, $LiCF_3SO_3$, $LiBF_4$, and dimethyl phthalate (DMP) as the plasticizer.

Hydroxyethyl cellulose (HEC) is a very important cellulose derivative due to its good viscosimetric, emulsifying, stabilizing, dispersing, and agglutinant properties as well as its high solubility in several organic solvents and water and its ability to form films with desirable mechanical properties (Fig. 3.45).

These characteristics make HEC suitable for applications in ink, paper, food, pharmaceutical, cosmetic, ceramic, textile, and agricultural products. Machado et al. [80] studied HEC with different quantities of glycerol and the addition of lithium trifluoromethane sulfonate ($LiCF_3SO_3$) salt conductivity values of $1.07 \times 10^{-5}\,S\,cm^{-1}$ at 30°C. It is a great challenge to prepare proton conducting HEC because acid hydrolysis of intra- and intermolecular hydrogen bonding takes place, thereby losing its film forming properties. Moreover, HEC is crystalline in nature and the addition of plasticizers, cross-linkers, and different dopants certainly improves the ionic conducting and stability of the polymer electrolyte. It is well known that borax (or sodium tetraborate, $Na_2B_4O_7$) is a good buffer and cross-linker. At lower concentrations, borax totally dissociates into an equivalent amount of boric acid and borate ion [81]. Usually borate ions and polymer molecules form complexes that induce electrostatic charges on the polymeric chains. The Na^+ ions will have a shielding effect on the charges of polymer chains. [82]. In addition, the hydroxyl groups of HEC form hydrogen bonding with borate ions, thus stabilizing the polymer matrix and creating the opportunity to dope acid into the HEC system.

Cellulose acetate propionate (CAP) is a mixed cellulose ester developed for coating applications. It is soluble in both organic solvents and in alcohol/water mixtures, has a fast solvent release rate, and is the basis of an excellent film-forming polymer (Fig. 3.46).

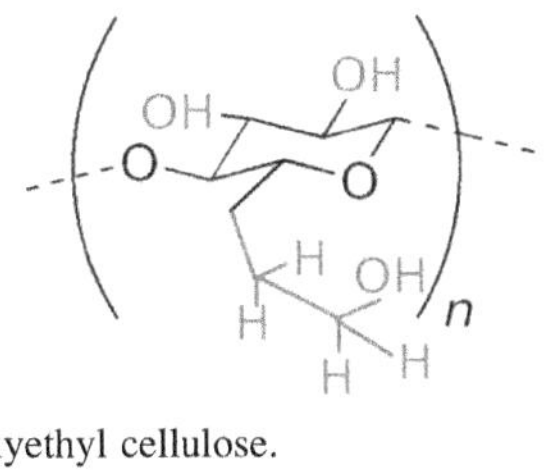

FIG. 3.45 Structure of hydroxyethyl cellulose.

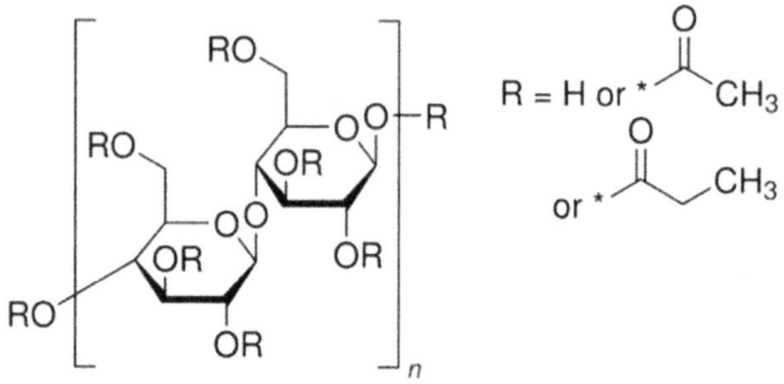

FIG. 3.46 Structure of cellulose acetate propionate.

A Case Study on the Preparation and Characterization of Phosphoric Acid-Doped Hydroxyethyl Cellulose Electrolyte for Use in a Supercapacitor

Experimental

Materials Preparation HEC (Aldrich, average M_w ~250,000) and borax (Aldrich, 99% greener alternative product) were mixed in 4:1 wt.% ratio taken in different petri plates. Twenty milliliter of Millipore water was added to the mixture and stirred well to form a uniform viscous solution. 0.1, 0.2, 0.4, and 0.5 mL of 85% of H_3PO_4 was added with stirring and allowed to settle for 2 h and subjected to 60°C for 2 h. SPE films obtained were peeled out and subjected to further studies.

Characterization FTIR measurements of the SPEs were carried out at room temperature using a Shimadzu 8400S FTIR spectrometer. Differential scanning calorimetery (DSC) measurements were done using a DSC-60, Shimadzu. Measurements were performed over a temperature range of 303–473 K at a heating rate of 10° C min^{-1} under the nitrogen atmosphere at a flow rate of 50 mL min^{-1}. Readings were taken from the first heating run.

For electrochemical studies, samples having a thickness of ~1 mm were cut into 1 cm × 1 cm square dimensions and placed between two square stainless steel electrodes (length 1 cm) fitted with copper wires. The whole setup was held tightly with a plastic clamp. The bulk ionic conductivities (σ) and dielectric properties of the SPEs were determined using the EIS in the frequency range of 1–100 MHz using a small amplitude AC signal of 10 mV. Experiments were carried out in the temperature range of 303–333 K using a PID-controlled oven from SES instruments Pvt. Ltd. From these EIS data, dielectric studies were also carried out.

Fabrication of a Symmetrical Supercapacitor Cell The electrode material for supercapacitor fabrication was prepared using activated carbon (AC) derived from areca fibers. AC was mixed with PVdF as a binder (PVdF/AC in 0.3:1 weight ratio) to form a slurry. This slurry was screen printed on two stainless steel electrodes. The supercapacitor cell was constructed using SPE sandwiched between two prepared AC-coated electrodes. The unit cell was sealed in a plastic coated aluminum pouch, keeping the two wires outside. Electrochemical characterization was carried out by CV, EIS, and GCD studies.

Results and Discussion

The bulk ionic conductivities (σ) of the SPEs have been determined from the complex impedance. The temperature-dependent ionic conductivity measurements were considered in order to analyze the possible mechanism of ionic conduction in these systems. Fig. 3.47 reveals Arrhenius plots for five concentrations of H_3PO_4 in the HEC electrolytes. Such behavior is observed in all the characterized polymer electrolytes, meaning that there is neither phase transition in the polymer matrix nor a domain formed by the addition of H_3PO_4.

The increase in the conductivity with the temperature can be interpreted as a hopping mechanism between coordinating sites, hopping being assisted by local structural relaxations and segmental motions of the polymer borax complexes. The conductivity values are found in the range of 10^{-3}–$10^{-4}\,S\,cm^{-1}$ for SPEs with different concentrations at different temperatures. The highest ionic conductivity value was $6.2 \times 10^{-3}\,S\,cm^{-1}$ at 343 K and $4.1 \times 10^{-3}\,S\,cm^{-1}$ at 303 K. The effective bonding of borax with the HEC would have helped to retain the acidic protons and hence exhibited stability during varying the temperature. Fig. 3.48 shows the Nyquist plot of the sample showing the highest conductivity for the calculation of the bulk resistance from the high-frequency intercept on the real impedance axis as a function of temperature.

As observed, the resistance of the films decreases as the temperature increases. At higher temperatures, thermal movement of polymer chain segments and the dissociation of acid would be improved, inducing an increase in total ionic conductivity. At a higher-frequency region, semicircles were observed indicating polarization of ions near the electrode/electrolyte interface region. At a low-frequency region, the similar vertical lines indicating the temperature effect on the diffusion of ions toward the electrode/electrolyte region is equally distributed in the polymer matrix.

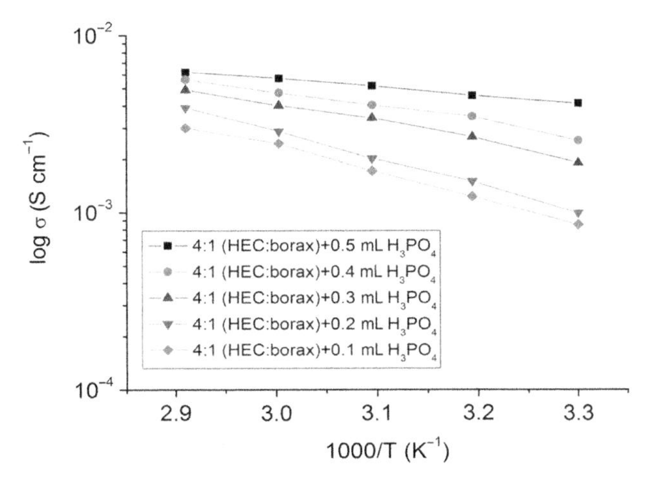

FIG. 3.47 Variations of bulk conductivities of SPEs with different temperatures.

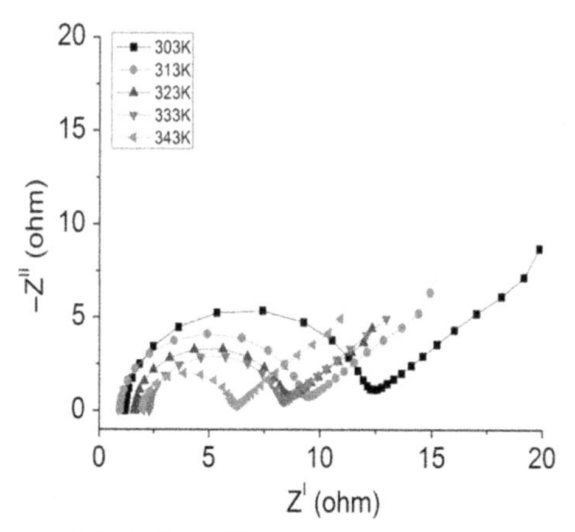

FIG. 3.48 Nyquist plots of the SPEs at different temperatures.

Fig. 3.49 shows activation energy, E_A, which was opposite to the conductivity. The values of E_A for SPEs studied in this work are in the range of 0.092–0.12 eV. The electrolyte with the lower value of E_A implies that dopant acid has been dissociated, favoring proton movement by forming coordination with other polymer sites. Comparatively, the present ionic conductivity is better than other ionic conductivities exhibited by natural polymer electrolytes, which ranged from 10^{-3} to 10^{-6} S cm^{-1} [83,84].

The dielectric constant (ε_R) is a measure of stored charge. The variation of dielectric constant, dielectric loss (ε_I), and the real (M_R) and imaginary (M_I) parts of the electrical modulus as a function of temperature are shown in Fig. 3.50. There are no appreciable relaxation peaks observed in the frequency range used in the study. Both ε_R and ε_I increase sharply at low frequencies, indicating that electrode polarization and space charge effects have occurred, confirming non-Debye dependence. On the other hand, at high frequencies, periodic reversal of the electric field occurs so fast that there is no excess ion diffusion in the direction of the field. Polarization due to charge accumulation decreases, which leads to the decrease in ε_R and ε_I. The ε_R and ε_I increase at higher temperatures because of the higher charge carrier density.

In the ε_I spectrum, no significant relaxation peaks have been observed in Fig. 3.50. This is indicates that residual water does not contribute toward conductivity enhancement. Both M_R and M_I show an increase at the higher frequency end and exhibit a long tail feature at the low frequency end. This indicates that the material is very capacitive in nature. M_I shows relaxation peaks at the lower-frequency region. This may be attributed to an increase in chaotic thermal oscillations of unbound HEC molecules and increased disorientation of dipoles.

The FTIR spectrum of HEC (Fig. 3.51) shows bands at 3487, 2923, 1658, and 1064 cm^{-1} characteristic for hydrogen stretching of bonded O—H, hydrogen stretching of C—H, double bond stretching of carbonyl, carboxyl, and carboxyl salt, and hydrogen bending of O—H, respectively.

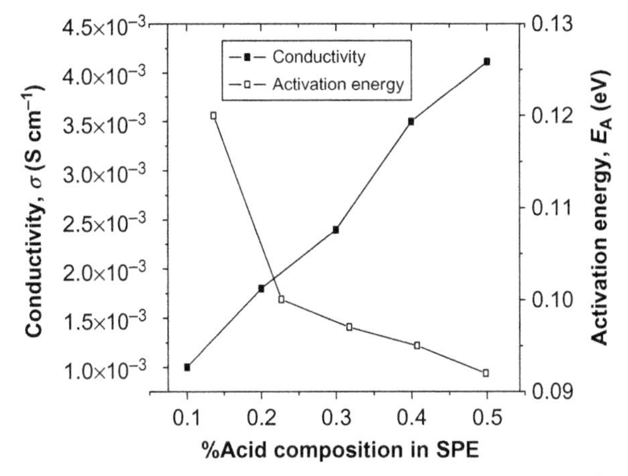

FIG. 3.49 Variation of activation energy (E_A) and ionic conductivity (σ) as a function of acid.

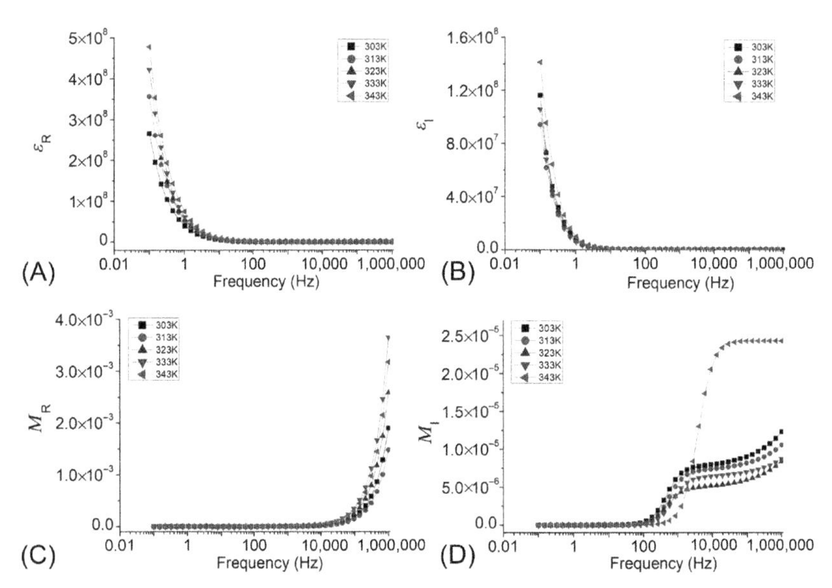

FIG. 3.50 Plots of dielectric studies versus frequency at different temperatures (A) dielectric constant, (B) dielectric loss, (C) real part of the electric modulus, and (D) imaginary part of the electric modulus.

The acid doped HEC shows peaks at 887–1303 cm^{-1}, which are attributed to free H_3PO_4 molecules and its anions. Peaks of carboxylate groups shifted to 1674 and 1465 cm^{-1} [85]. $H_2PO_4^-$ peak at 887 cm^{-1} was observed. Borax peaks were also observed at 702 and 440 cm^{-1} were of the O—B—O group [86]. It indicates that H_3PO_4 was able to hydrolyze the hydrogen stretching of O—H as

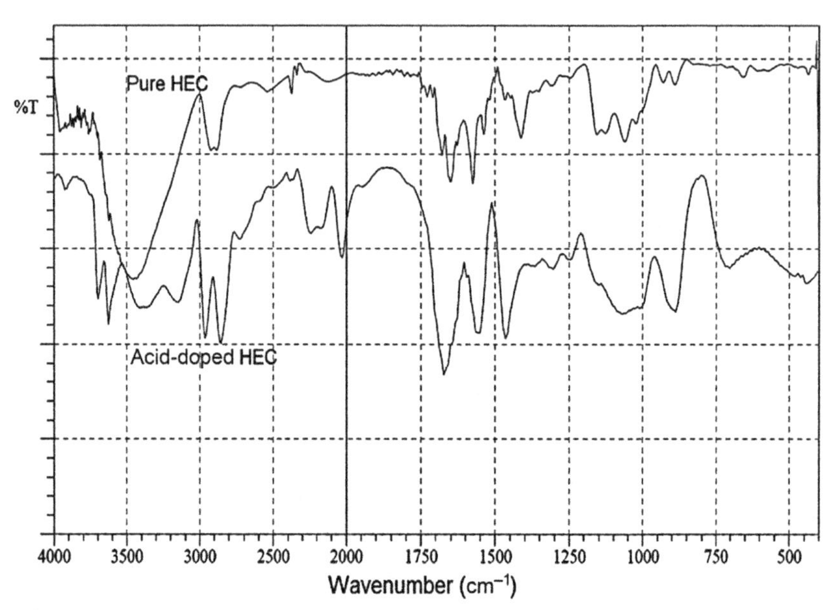

FIG. 3.51 FTIR spectra of pure HEC and H_3PO_4-doped HEC.

observed 2854–3363 cm^{-1} regions, but the presence of borax helped to maintain the stability of the polymer matrix.

Fig. 3.52 shows DSC thermograms of pure HEC and acid-doped HEC. A small shoulder peak was observed in pure HEC at 135°C, which is the glass transition temperature (T_g) of the HEC.

The second thermogram shows a shift in the T_g value (130°C). This fact is evidenced by the tendency of the T_g to decrease in films with higher acid content. Both acid and water (naturally absorbed by the electrolyte) are small molecules that lodge among the polymeric chains, promoting an increase in the motion of their segments in the amorphous phase. It reduces the intermolecular attraction forces due to the breaking of hydrogen bonds among hydroxyl groups belonging to different chains. Consequently, the system becomes more flexible, which results in a lower T_g value. An exothermic peak was observed for both the samples at higher temperatures, indicating the thermal degradation of the samples starts at about 220°C and continues to 310°C over which no more changes are observed. This indicates a formation of solid residues due to the degradation product of polymer and borax.

Supercapacitor Studies CV responses for the carbon-carbon symmetrical supercapacitor at various sweep rates are shown in Fig. 3.53.

A maximum specific capacitance of 83 $F g^{-1}$ at a scan rate of 2 $mV s^{-1}$ was obtained, implying that a large amount of mobile protons are available in the film, which leads to a high capacitance of the device. The behaviors observed are characteristic of double-layer capacitive features. In all CVs, the increase in area under the voltammograms was observed without any major distortion of the rectangular

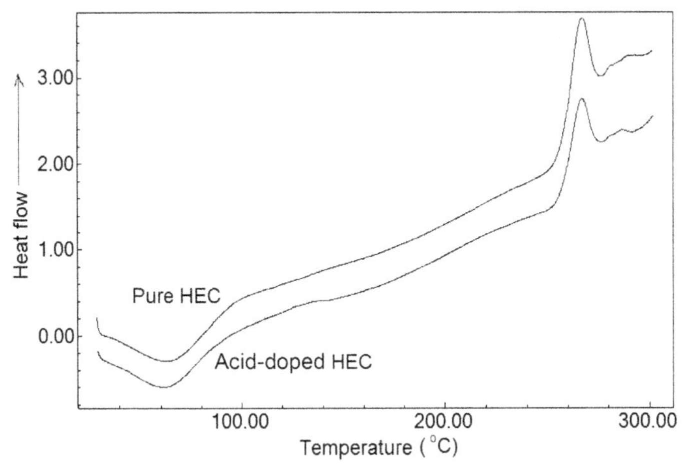

FIG. 3.52 DSC thermograms of HEC and acid-doped HEC.

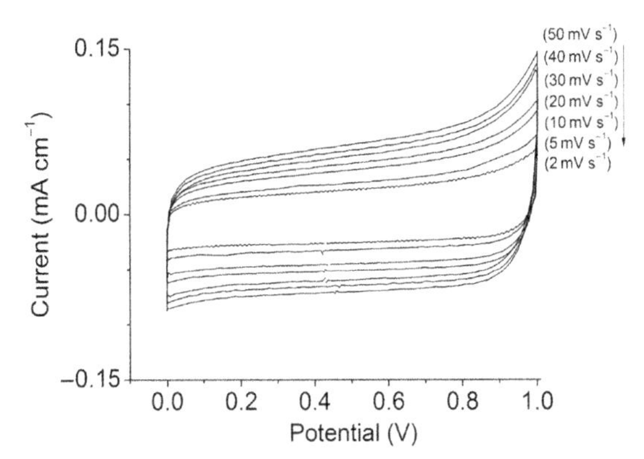

FIG. 3.53 CVs of the supercapacitor at different scan rates.

window, indicating that these electrodes exhibit typical capacitive behavior with good reversibility and high stability for scan rates from 2 to 50 mV s^{-1}. A slight deviation from such a rectangular shape indicates low equivalent series resistance (ESR) in the device. Table 3.3 shows a comparison with other similar electrode materials and electrolytes.

The AC impedance response (Nyquist plot) of the carbon-carbon supercapacitor is shown in Fig. 3.54. The plot shows a semicircle at a high-frequency range and a straight line in the low-frequency region.

Using this resistance at high frequency, the value of C_{dl}, the double-layer capacitance, has been determined from the high-frequency region of the impedance spectrum; the C_{dl} value obtained was 5 mF cm^{-2}. The capacitance value increases at low frequencies because of a large number of ionic movements. The semicircle

TABLE 3.3 Comparison of Specific Capacitance With Other Electrode Materials and Electrolytes

Electrode Material	Electrolyte	Specific Capacitance	Ref.
NiO and activated carbon	PVA-KOH-H_2O	73.4 Fg^{-1}	[87]
Black pearl carbon	Gelatin + NaCl	81 Fg^{-1}	[88]
Activated carbon (BP20)	Methyl cellulose + NH_4NO_3	39 Fg^{-1}	[89]
Activated carbon	HEC + H_3PO_4	83 Fg^{-1}	Present work

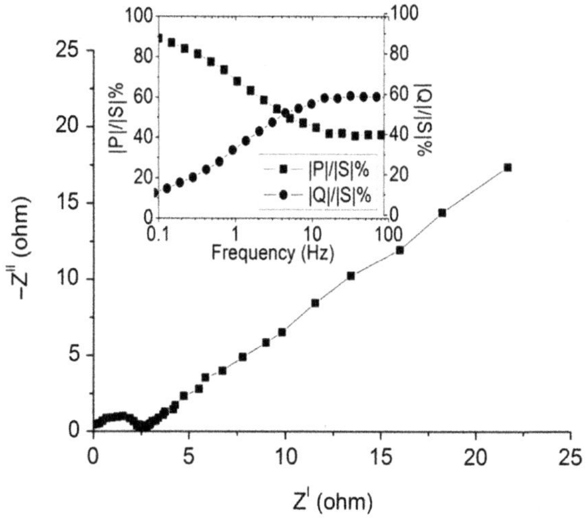

FIG. 3.54 AC impedance plot of the supercapacitor (inset: plots of normalized reactive power |Q|/|S|% and active power |P|/|S|% versus frequency (Hz)).

results from the parallel combination of resistance and capacitance and the linear region is because of the Warburg impedance.

Using normalized reactive power |Q|/|S|% and active power |P|/|S|% versus the frequency plot for the 1 cm^2 cell, the time constant of the fabricated supercapacitor has been calculated and is shown in Fig. 3.54 (inset). The calculated time constant was found to be equal to 0.2 s.

Fig. 3.55 represents the charge-discharge curves at different current density, namely 0.1, 0.2, 0.3, and 0.4 Ag^{-1} at a potential range of 0–1 V. It is apparent from

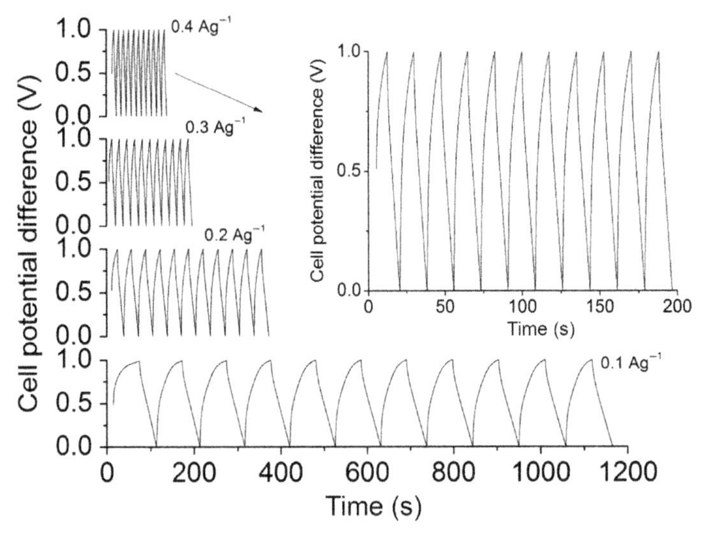

FIG. 3.55 Galvanostatic charge/discharge plots of the supercapacitor at different current densities.

Fig. 3.8 that the charging time and the discharging pattern remained constant with an increasing number of cycles, but it decreased with an increase in current density.

The voltage-time responses behaving like a mirror during the charge-discharge process meant that the SC owned a good electrochemical capacitance performance. The specific capacitance was found to be 110 F g^{-1} for the first cycle with a coulombic efficiency of 98%, which is found to be similar to the capacitance value from CV measurements. Equivalent series resistance (ESR) was 24 Ω for the first cycle with a voltage drop of 0.03 V. Specific energy (SE) and specific power (SP) values ranged between 4 Wh kg^{-1} and 830 W kg^{-1}. Fig. 3.56 shows cyclic

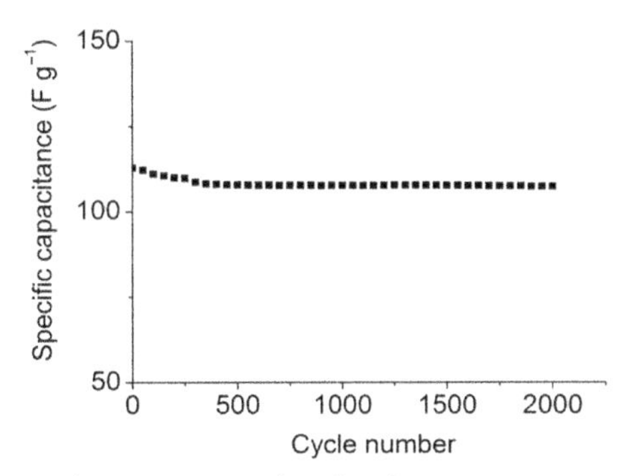

FIG. 3.56 Specific capacitance versus number of cycles.

stability during charge–discharge studies up to 2000 cycles at $0.2\,A\,g^{-1}$. The stability of the supercapacitor material was quite good as seen in the capacitance values in the plot. The initial decrease in the capacitance values is minimum, which is due to the loss of the charges initially stored at interfaces associated with the irreversible reactions of loosely bound surface groups on the porous activated carbon electrodes.

(Sudhakar YN, Selvakumar M, Bhat DK. Preparation and characterization of phosphoric acid-doped hydroxyethyl cellulose electrolyte for use in supercapacitor. Mater Renew Sustain Energy 2015;4:10.)

3.5 CONCLUSION

SPE film was prepared mixing HEC and borax at a 4:1 wt.% ratio and 0.5 mL 85% of H_3PO_4. By adding borax, the stability increased in the polymer chains in the presence of doped H_3PO_4. The highest ionic conductivity value was $6.2 \times 10^{-3}\,S\,cm^{-1}$ at 343 K and $4.1 \times 10^{-3}\,S\,cm^{-1}$ at 303 K. A better dissociation property of the SPE system was evident from the activation energy of 0.12 eV at 303 K. The dielectric studies indicated the high capacitive nature and supported the structure of SPE. FTIR and DSC characterization showed good interaction between the polymer and borax, thereby enhancing the free proton in the polymer matrix. A fabricated carbon-carbon supercapacitor showed specific capacitance of $83\,F\,g^{-1}$ at $2\,mV\,s^{-1}$. SE and SP values ranged between $4\,Wh\,kg^{-1}$ and $830\,W\,kg^{-1}$. Cyclic performance was quite stable during charge-discharge cycles, maintaining 98% coulombic efficiency.

REFERENCES

[1] Simon P, Gogotsi Y. Materials for electrochemical capacitors. Nat Mater 2008;7:845–54.
[2] Becker HI. Low voltage electrolytic capacitor. US Patent 2,800,616; 1957.
[3] Shi S, Xu C, Yang C, Li J, Du H, Li B, et al. Flexible supercapacitors. Particuology 2013;371–7.
[4] Niu Z, Zhang L, Liu L, Zhu B, Dong H, Chen X. All-solid-state flexible ultrathin micro-supercapacitors based on graphene. Adv Mater 2013;25:4035–42.
[5] John RM, Patrice S. Fundmentals of electrochemical capacitor design and operation. Electrochem Soc Interface 2008;Spring.
[6] Wang G, Zhang L, Zhang J. A review of electrode materials for electrochemical supercapacitors. Chem Soc Rev 2012;41:797–828.
[7] Rufford TE, Hulicova-Jurcakova D, Zhu Z, Lu GQ. Nanoporous carbon electrode from waste from waste coffee beans for high performance supercapacitors. Electrochem Commun 2008;10:1594–7.
[8] Gamby J, Taberna P, Simon P, Fauvarque J, Chesneau M. Studies and characterisations of various activated carbons used for carbon/carbon supercapacitors. J Power Sources 2001;101:109–16.
[9] Béguin F, Presser V, Balducci A, Frackowiak E. Carbons and electrolytes for advanced supercapacitors. Adv Mater 2014;26:1–33.

[10] Conway BE. Electrochemical supercapacitors: scientific fundamentals and technological applications. New York: Kluwer Academic/Plenum Publisher; 1999.
[11] Zhang LL, Zhao XS. Carbon-based materials as supercapacitor electrodes. Chem Soc Rev 2009;38(9):2520–31.
[12] Eigler S, Enzelberger-Heim M, Grimm S, Hofmann P, Kroener W, Geworski A, et al. Wet chemical synthesis of graphene. Adv Mater 2013;25(26):3583–7.
[13] Tang L, Wang Y, Li Y, Feng H, Lu J, Li J. Preparation, structure, and electrochemical properties of reduced graphene sheet films. Adv Funct Mater 2009;19(17):2782–9.
[14] Chen H, Müller MB, Gilmore KJ, Wallace GG, Li D. Mechanically strong, electrically conductive, and biocompatible graphene paper. Adv Mater 2008;20(18):3557–61.
[15] Cai M, Thorpe D, Adamson DH, Schniepp HC. Methods of graphite exfoliation. J Mater Chem 2012;24992–5002.
[16] Fu Y, Zhang J, Liu H, Hiscox WC, Gu Y, Mermin ND, et al. Ionic liquid-assisted exfoliation of graphite oxide for simultaneous reduction and functionalization to graphenes with improved properties. J Mater Chem A 2013;1(7):2663.
[17] Chabot V, Kim B, Sloper B, Tzoganakis C, Yu A. High yield production and purification of few layer graphene by gum arabic assisted physical sonication. Sci Rep 2013;3:1378.
[18] Bo Z, Shuai X, Mao S, Yang H, Qian J, Chen J, et al. Green preparation of reduced graphene oxide for sensing and energy storage applications. Sci Rep 2014;4:4684.
[19] Zhang D, Zhang X, Chen Y, Wang C, Ma Y. An environment-friendly route to synthesize reduced graphene oxide as a supercapacitor electrode material. Electrochim Acta 2012;69:364–70.
[20] Jin Y, Huang S, Zhang M, Jia M, Hu D. A green and efficient method to produce graphene for electrochemical capacitors from graphene oxide using sodium carbonate as a reducing agent. Appl Surf Sci 2013;268:541–6.
[21] Sun Y, Wu Q, Shi G. Graphene based new energy materials. Energy Environ Sci 2011; 4(4):1113–32.
[22] Hantel MM, Kaspar T, Nesper R, Wokaun A, Kötz R. Partially reduced graphite oxide for supercapacitor electrodes: effect of graphene layer spacing and huge specific capacitance. Electrochem Commun 2011;13(1):90–2.
[23] Li ZJ, Yang BC, Zhang SR, Zhao CM. Graphene oxide with improved electrical conductivity for supercapacitor electrodes. Appl Surf Sci 2012;258(8):3726–31.
[24] Murugan AV, Muraliganth T, Manthiram A. Rapid, facile microwave-solvothermal synthesis of graphene nanosheets and their polyaniline nanocomposites for energy storage. Chem Mater 2009;21:2692.
[25] Zhang L, Zhang F, Yang X, Long G, Wu Y, Zhang T, et al. Porous 3D graphene-based bulk materials with exceptional high surface area and excellent conductivity for supercapacitors. Sci Rep 2013;3:1408.
[26] Zhou YK, He BL, Zhou WJ, Huang J, Li XH, Wu B, et al. Electrochemical capacitance of well-coated single-walled carbon nanotube with polyaniline composites. Electrochim Acta 2004; 49(2):257–62.
[27] Rudge A, Davey J, Raistrick I, Gottesfeld S, Ferraris JP. Conducting polymers as active materials in electrochemical capacitors. J Power Sources 1994;47(1–2):89–107.
[28] Zang J, Bao SJ, Li CM, Bian H, Cui X, Bao Q, et al. Well-aligned cone-shaped nanostructure of polypyrrole/RuO2 and its electrochemical supercapacitor. J Phys Chem C 2008; 112(38):14843–7.
[29] Taberna PL, Simon P, Fauvarque JF. Electrochemical characteristics and impedance spectroscopy studies of carbon-carbon supercapacitors. J Electrochem Soc 2003;150(3):A292.

[30] Kadir MFZ, Aspanut Z, Majid SR, Arof AK. FTIR studies of plasticized poly(vinyl alcohol)–chitosan blend doped with NH_4NO_3 polymer electrolyte membrane. Spectrochim Acta A Mol Biomol Spectrosc 2011;78:1068–74.
[31] Shukur MF, Ithnin R, Illias HA, Kadir MFZ. Proton conducting polymer electrolyte based on plasticized chitosan-PEO blend and application in electrochemical devices. Opt Mater 2013;35:1834–41.
[32] Sousa AMM, Sereno AM, Hilliou L, Gonçalves MP. Biodegradable agar extracted from *Gracilaria Vermiculophylla*: film properties and application to edible coating. Mater Sci Forum 2010;636–637:739–44.
[33] Polu AR, Kumar R. AC impedance and dielectric spectroscopic studies of Mg^{2+} ion conducting PVA—PEG blended polymer electrolytes. Bull Mater Sci 2011;34:1063–7.
[34] Kolhe P, Kannan RM. Improvement in ductility of chitosan through blending and copolymerization with PEG: FTIR investigation of molecular interactions. Biomacromolecules 2003;4(1):173–80.
[35] Rajendran S, Babu R, Renuka K. Ionic conduction behavior in PVC—PEG blend polymer electrolytes upon the addition of TiO_2. Ionics 2009;15:61–6.
[36] Sun J, Jordan LR, Forsyth M, MacFarlane DR. Acid-organic base swollen polymer membranes. Electrochim Acta 2001;46(10 – 11):1703–8.
[37] Acar O, Sen U, Bozkurt A, Ata A. Proton conducting membranes based on poly(2,5-benzimidazole) (ABPBI)-poly(vinylphosphonic acid) blends for fuel cells. Int J Hydrog Energy 2009;34(6):2724–30.
[38] Bruce PG, Vincent CA. Steady state current flow in solid binary electrolyte cells. J Electroanal Chem Interfacial Electrochem 1987;225:1.
[39] Evans J, Vincent CA, Bruce PG. Electrochemical measurement of transference numbers in polymer electrolytes. Polymer (Guildf) 1987;28(13):2324–8.
[40] Osman Z, Ibrahim ZA, Arof AK. Conductivity enhancement due to ion dissociation in plasticized chitosan based polymer electrolytes. Carbohydr Polym 2001;44(2):167–73.
[41] Subba Reddy C, Sharma A, Narasimha Rao VV. Conductivity and discharge characteristics of polyblend (PVP+PVA+KIO3) electrolyte. J Power Sources 2003;114(2):338–45.
[42] Ramesh S, Yahaya AH, Arof AK. Dielectric behaviour of PVC-based polymer electrolytes. Solid State Ionics 2002;152–153:291–4.
[43] Qian X, Gu N, Cheng Z, Yang X, Wang E, Dong S. Impedance study of $(PEO)_{10}LiClO_4$-Al_2O_3 composite polymer electrolyte with blocking electrodes. Electrochim Acta 2001; 46(12):1829–36.
[44] Forsyth M, MacFarlane DR, Meakin P, Smith ME, Bastow TJ. An NMR investigation of ionic structure and mobility in plasticized solid polymer electrolytes. Electrochim Acta 1995; 40:2343–7.
[45] Chowdari BVR, Chandra S, Singh S, Srivastava PC, editors. Solid state ionics: materials and applications. Singapore: World Scientific; 1992. p. 373.
[46] Ling-hao H, Rui X, De-bin Y, Ying L, Rui S. Effects of blending chitosan with PEG on surface morphology, crystallization and thermal properties. Chinese J Poly Sci 2009;27(4): 501–10.
[47] Choudhury NA, Shukla AK, Sampath S, Pitchumani S. Cross-linked polymer hydrogel electrolytes for electrochemical capacitors. J Electrochem Soc 2006;153(3):A614.
[48] Stephan AM. Review on gel polymer electrolytes for lithium batteries. Eur Polym J 2006;42:21–42.
[49] Baskakova YV, Yarmolenko OV, Efimov ON. Polymer gel electrolytes for lithium batteries. Russ Chem Rev 2012;81(4):367–80.

[50] Barbucci R, Pasqui D, Favaloro R, Panariello G. A thixotropic hydrogel from chemically cross-linked guar gum: synthesis, characterization and rheological behaviour. Carbohydr Res 2008;343(18):3058–65.
[51] Sandolo C, Matricardi P, Alhaique F, Coviello T. Effect of temperature and cross-linking density on rheology of chemical cross-linked guar gum at the gel point. Food Hydrocoll 2009;23 (1):210–20.
[52] da Silva GP, Mack M, Contiero J. Glycerol: a promising and abundant carbon source for industrial microbiology. Biotechnol Adv 2009;27(1):30–9.
[53] Mattos RI, Raphael E, Majid SR, Arof AK, Pawlicka A. Enhancement of electrical conductivity in plasticized chitosan based membranes. Mol Cryst Liq Cryst 2012;554(1):150–9.
[54] Matricardi P, Cencetti C, Ria R, Alhaique F, Coviello T. Preparation and characterization of novel gellan gum hydrogels suitable for modified drug release. Molecules 2009; 14(9):3376–91.
[55] Fialho AM, Moreira LM, Granja AT, Popescu AO, Hoffmann K, Sá-Correia I. Occurrence, production, and applications of gellan: current state and perspectives. Appl Microbiol Biotechnol 2008;79(6):889–900.
[56] Petri DFS. Xanthan gum: a versatile biopolymer for biomedical and technological applications. J Appl Polym Sci 2015;132(23).
[57] Song MK, Kim YT, Kim YT, Cho BW, Popov BN, Rhee HW. Thermally stable gel polymer electrolytes. J Electrochem Soc 2003;150(4):A439–44.
[58] Raducha D, Wieczorek W, Florjanczyk Z, Stevens JR. Nonaqueous H3PO4-doped gel electrolytes. J Chem Phys 1996;3654(96):20126–33.
[59] Ponrouch A, Garbarino S, Bertin E, Guay D. Ultra high capacitance values of Pt@RuO 2 core-shell nanotubular electrodes for microsupercapacitor applications. J Power Sources 2013;221:228–31.
[60] Choudhury NA, Northrop PWC, Crothers AC, Jain S, Subramanian VR. Chitosan hydrogel-based electrode binder and electrolyte membrane for EDLCs: experimental studies and model validation. J Appl Electrochem 2012;42:935–43.
[61] Jung HG, Venugopal N, Scrosati B, Suna YK. A high energy and power density hybrid supercapacitor based on an advanced carbon-coated $Li_4Ti_5O_{12}$ electrode. J Power Sources 2013;221:266–71.
[62] Du Pasquier A, Laforgue A, Simon P, Amatucci GG, Fauvarque JF. A nonaqueous asymmetric hybrid $Li_4Ti_5O_{12}$/poly(fluorophenylthiophene) energy storage device. J Electrochem Soc 2002;149(3):A302–6.
[63] Yu H, Wu J, Fan L, Xu K, Zhong X, Lin Y, et al. Improvement of the performance for quasi-solid-state supercapacitor by using PVA–KOH–KI polymer gel electrolyte. Electrochim Acta 2011;56(20):6881–6.
[64] Irimia-Vladu M. Green electronics: Biodegradable and biocompatible materials and devices for sustainable future. Chem Soc Rev 2014;43:588.
[65] Varshney P, Gupta S. Natural polymer-based electrolytes for electrochemical devices: a review. Ionics 2011;17:479–83.
[66] Murata K, Izuchi S, Yoshihisa Y. An overview of the research and development of solid polymer electrolyte batteries. Electrochim Acta 2000;45(8–9):1501–8.
[67] Senthilkumar ST, Selvan RK, Ponpandian N, Melo JS. Redox additive aqueous polymer gel electrolyte for an electric double-layer capacitor. RSC Adv 2012;2(24):8937–40.
[68] Wright PV. Polymer electrolytes-the early days. Electrochim Acta 1998;43:1137.
[69] Sudhakar YN, Selvakumar M. Lithium perchlorate doped plasticized chitosan and starch blend as biodegradable polymer electrolyte for supercapacitors. Electrochim Acta 2012;78:398–405.

[70] Arof AK, Morni NM. Chitosan-Lithium triflate electrolyte in secondary lithium cells. J Power Sources 1999;77:42–8.
[71] Osman Z, Arof AK. FTIR studies of chitosan acetate based polymer electrolytes. Electrochim Acta 2003;48:993–9.
[72] Idris NK, Aziz NAN, Zambri MSM, Zakaria NA, Isa MIN. Ionic conductivity studies of chitosan-based polymer electrolytes doped with adipic acid. Ionics (Kiel) 2009;15:643–6.
[73] Bhatt AS, Bhat DK, Santosh MS. Electrochemical properties of chitosan—Co_3O_4 nanocomposite films. J Electroanal Chem 2011;657:135–43.
[74] Mohamed NS, Subban RHY, Arof AK. Polymer batteries fabricated from lithium complexed acetylated chitosan. J Power Sources 1995;56:153.
[75] Yahya MZA, Arof AK. Effect of oleic acid plasticizer on chitosan–lithium acetate solid polymer electrolytes. Eur Polym J 2003;39:897.
[76] Wee G, Larsson O, Madhavi S, Magnus B, Xavier C, Subodh M. Effect of the ionic conductivity on the performance of polyelectrolyte-based supercapacitors. Adv Funct Mater 2010;20:4344.
[77] Mohamad AA, Mohamed NS, Yahya MZA, Othman R, Ramesh S, Alias Y, et al. Ionic conductivity studies of poly(vinyl alcohol) alkaline solid polymer electrolyte and its use in nickel-zinc cells. Solid State Ionics 2003;156(1–2):171–7.
[78] Fonseca CP, Neves S. The usefulness of a LiMn2O4 composite as an active cathode material in lithium batteries. J Power Sources 2004;135(1–2):249–54.
[79] Rajendran S, Sivakumar M, Subadevi R. Li-ion conduction of plasticized PVA solid polymer electrolytes complexed with various lithium salts. Solid State Ionics 2004;167(3–4):335–9.
[80] MacHado GO, Ferreira HCA, Pawlicka A. Influence of plasticizer contents on the properties of HEC-based solid polymeric electrolytes. Electrochim Acta 2005;50(19):3827–31.
[81] Pezron E, Ricard A, Lafuma T, Audebert R. Reversible gel formation induced by ion complexation. 1. Borax-galactomannan interactions. Macromolecules 1988;21:1121.
[82] Pezron E, Leibler L, Ricard A, Lafuma T, Audebert R. Reversible gel formation induced by ion complexation. 2. Phase diagrams. Macromolecules 1988;21:1126.
[83] Pawlicka M, Danczuk W, Wieczorek E. Zygadlo-Monikowska. Influence of plasticizer type on the properties of polymer electrolytes based on chitosan. J Phys Chem A 2008;112:8888.
[84] Mohamed S, Johari N, Ali A. Electrochemical studies on epoxidised natural rubber-based gel polymer electrolytes for lithium–air cells. J Power Sources 2008;183:351–4.
[85] Arof AK, Majid SR. FTIR studies of chitosan-orthophosphoric acid-ammonium nitrate-aluminosilicate polymer. Mol Cryst Liq Cryst 2008;484:107.
[86] Judeninstein T, Reichert P, de Azevedo D, Bonagambac ER. NMR multi-scale description of ionic conductivity mechanisms. Acta Chim Slov 2005;52:349.
[87] Zhang LL, Zhao X, Stoller MD, Zhu Y, Ji H, Murali S, et al. Highly conductive and porous activated reduced graphene oxide films for high-power supercapacitors. Nano Lett 2012;12(4):1806–12.
[88] Yuan C, Zhang X, Wu Q, Gao B. Effect of temperature on the hybrid supercapacitor based on NiO and activated carbon with alkaline polymer gel electrolyte. Solid State Ionics 2006;177(13–14):1237–42.
[89] Shuhaimi NEA, Teo LP, Woo HJ, Majid SR, Arof AK. Electrical double-layer capacitors with plasticized polymer electrolyte based on methyl cellulose. Polym Bull 2012;69:807–26.

Chapter 4

Biopolymer Electrolytes for Solar Cells and Electrochemical Cells

Chapter Outline

Biopolymer Electrolytes. https://doi.org/10.1016/B978-0-12-813447-4.00004-2

4.1 PHOTONS IN, ELECTRONS OUT: THE PHOTOVOLTAIC EFFECT

Solar photovoltaic energy conversion is a one-step conversion process that generates electrical energy from light energy. Light is made up of packets of energy, called photons, whose energy depends only upon the frequency, or color, of the light. The energy of visible photons is sufficient to excite electrons, bound into solids, up to higher energy levels where they are more free to move. Normally, when light is absorbed by matter, photons are given up to excite electrons to higher energy states within the material, but the excited electrons quickly relax back to their ground state (Fig. 4.1).

4.2 NEED OF THE SOLAR CELL

The energy sources that are either unlimited or can be quickly replaced and regenerated are called renewable energy sources. Though these sources have long been known, perhaps even before the start of civilization, their use on demand remains a big challenge. While most of these sources are still at the development stage, a few have reached a level of commercialization. However, the field still needs more research and development to produce and supply 20% of our electricity by 2020 [1,2]. In most countries, the major governmental emphasis has been focused toward development of renewable energy sources.

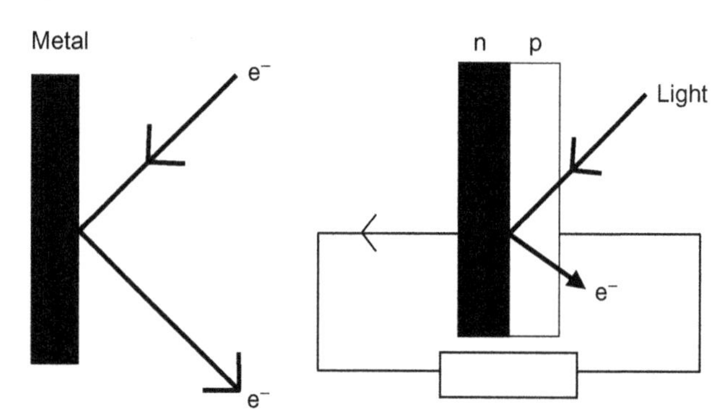

FIG. 4.1 Comparison of the photoelectric effect (left), where UV light liberates electrons from the surface of a metal, with the photovoltaic effect in a solar cell.

However, the awareness and interest toward the use of renewable resources need to be further enhanced. Of all popular renewable energy sources, solar energy is probably the most well-known and studied source of energy. The sun was known as an energy source even before the dawn of civilization [3,4].

4.3 HISTORY OF THE SOLAR CELL

The photovoltaic effect was first reported by Edmund Bequerel in 1839 when he observed that the action of light on a silver-coated platinum electrode immersed in electrolyte produced an electric current. Forty years later, the first solid-state photovoltaic devices were constructed by workers investigating the recently discovered photoconductivity of selenium. In 1876, William Adams and Richard Day found that a photocurrent could be produced in a sample of selenium when contacted by two platinum contacts. In 1894, Charles Fritts prepared what was probably the first large area solar cell by pressing a layer of selenium between gold and another metal. In the following years, photovoltaic effects were observed in copper-copper oxide thin film structures as well as in lead sulfide and thallium sulfide. These early cells were thin-film Schottky barrier devices, where a semitransparent layer of metal deposited on the semiconductor provided both the asymmetric electronic junction, which is necessary for photovoltaic action, and access to the junction for the incident light. In the 1950s, the development of silicon electronics followed the discovery of a way to manufacture p-n junctions in silicon. Naturally n-type silicon wafers developed a p-type skin when exposed to the gas boron trichloride. Part of the skin could be etched away to give access to the n-type layer beneath. These p-n junction structures produced much better rectifying action than Schottky barriers and better photovoltaic behavior. The first silicon solar cell was reported by Chapin, Fuller, and Pearson in 1954 and converted sunlight with an efficiency of 6 percent. Also in 1954, a cadmium sulfide p-n junction was produced with an efficiency of 6%. In the following years, studies of p-n junction photovoltaic devices in gallium arsenide, indium phosphide, and cadmium telluride were stimulated by theoretical work indicating that these materials would offer a higher efficiency. During the 1990s, interest in photovoltaics expanded along with growing awareness of the need to secure an electrical alternative to fossil fuels. In the later 1990s, the photovoltaic production expanded at a rate of 15%–25% per annum, driving a reduction in cost. Photovoltaics first become competitive in contexts where the conventional electricity supply was most expensive, for instance, in remote low-power applications such as navigation, telecommunications, and rural electrification as well as for enhancement of supply in grid-connected loads at peak use. As prices fall, new markets open up. An important example is building integrated photovoltaic applications where the cost of the photovoltaic system is offset by the savings in building materials.

4.4 TYPES OF SOLAR CELLS

The two main members of this group are dye-sensitized solar cells (DSSC) and organic photovoltaic cells (OPV). A dye-sensitized solar cell (DSSC) is composed of two electrodes, the anode and the cathode. These electrodes are made from a specific glass that has a transparent conductive oxide (TCO) coating on one side. Due to their high energy conversion efficiency, low cost, flexibility, and environmental friendliness, dye-sensitized solar cells based on nanocrystalline porous TiO_2 film have attracted both academic and commercial interest [5–7]. A typical DSSC consists of a photoelectrode based on a nanocrystalline titanium dioxide mesoporous layer sensitized with a dye, a platinum counter electrode, and an electrolyte solution between the electrodes. The principle of energy conversion is based on the injection of electrons from the photo-excited dye to the conduction band of the semiconductor. Photo-generated charges can diffuse in the porous network and can be collected at the electrode. The oxidized dye is reduced by a redox mediator (usually I^-/I_3^-) present in the electrolyte, which in turn is reduced at the counter-electrode, thus making the cell regenerative [8].

4.5 OPERATING PRINCIPLE OF THE ORGANIC PHOTOVOLTAIC CELL (OPV)

Organic solar cells are made of thin layers of organic materials with thickness in the 100 nm range. The motivation for using organic dyes is to replace the expensive silicon in conventional photovoltaics and to apply simple production techniques. Additionally, organic solar cells can be prepared on plastic foil, making them ideal candidates for flexible and portable systems. Organic solar cells basically comprise the following layers: first electrode, electron transport layer, photoactive layer, hole transport layer, and second electrode. In general, a solar cell absorbs light, separates the created electrons and holes from each other, then delivers electrical power at the contacts. The fundamental difference between the working principles of organic and inorganic solar cells is the direct generation of free charge carries in the inorganic solar cells. In organic materials the light absorption is followed by the creation of excitons with a typical binding energy of 0.3–0.5 eV.

Because the necessary electric field ($>10^6$ V cm^{-1}) to overcome this binding energy is not available in an organic solar cell, the excitons are usually separated at the interface between two different organic layers (heterojunction). The energy alignment of these two materials has to be optimized so that on the one hand the excitons are efficiently separated, but on the other hand no energy might be lost in this process.

4.6 ELECTRODE MATERIALS USED IN SOLAR CELLS

Transparent conducting films (TCFs) are required in traditional DSSCs as well as a host of other applications, including flat panel displays and touchscreens. Normally, indium tin oxide (ITO) and fluorine-doped tin oxide (FTO) will be used as the photoanode in DSSC. TiO_2 nanoparticles provide a high surface area film for increased dye loading while also acting as a photoanode material in the majority of cases of DSSC. Thus, many research groups have focused on the binary metal oxides photoanodes of TiO_2, ZnO, SnO_2, and Zn_2SnO_4. TiO_2 has the highest cell conversion efficiency among other metal oxides but the wide band gap energy (3.60 eV) and higher electronic mobility (250 $cm^2 v^{-1}$ - s^{-1}) of SnO_2 can be a promising alternative to the TiO_2 semiconductor for DSSCs. Recently, graphene-based materials, such as pristine graphene, graphene oxide, and reduced graphene oxide, have properties that are attractive for various components of the DSSC photoanode. Thus, we introduce applications of graphene-based materials in each part of the DSSC photoanode, the transparent conducting electrode, the sensitizing material, and the semiconducting layer. The incorporation of graphene in the photoanode enhances the fast electron transfer ability, Young's modulus and transparency. Doping or surface modifications of graphene nanosheets with other materials can also improve the photoanode and, thus the resulting cell performance is enhanced [9,10].

4.7 ELECTROLYTES IN DYE-SENSITIZED SOLAR CELLS

The electrolyte is one of the most crucial components in DSSCs, responsible for the inner charge carrier transport between electrodes. It continuously regenerates the dye and itself during DSSC operation. The electrolyte has great influence on the light-to-electric conversion efficiency and long-term stability of the devices. The efficiency of a DSSC device is determined by its photocurrent density (J_{sc}), photovoltage (V_{oc}), and fill factor (FF) [11]. All three parameters will be significantly affected by the electrolyte in DSSCs and by the interaction of the electrolyte with the electrode interfaces. For instance, J_{sc} can be affected by the transport of the redox couple components in the electrolyte. The FF can be affected by the diffusion of the charge carrier in the electrolyte and the charge transfer resistance on the electrolyte/electrode interface. The V_{oc} can be significantly affected by the redox potential of the electrolyte. Several aspects and importance of the electrolytes used in DSSCs are [12].

(a) The electrolytes must be able to transport the charge carriers between the photoanode and the counter electrode. After the dye injects electrons into the conduction band of TiO_2, the oxidized dye must be rapidly reduced to its ground state. Thus, the choice of the electrolyte should take into account the redox potential and the regeneration of the dye and itself.

(b) The electrolytes must guarantee fast diffusion of charge carriers (higher conductivity) and produce good interfacial contact with the mesoporous semiconductor and the counter electrode. For liquid electrolytes, the solvent should have smaller leakage and/or evaporation to prevent loss of the liquid electrolyte.

(c) The electrolytes must have long-term stabilities, including chemical, thermal, optical, electrochemical, and interfacial. They also must not cause desorption and degradation of the sensitized dye.

(d) The electrolytes should not exhibit a significant absorption in the range of visible light. Because the iodide/triiodide redox couple in the electrolyte shows color and reduces visible light absorption, triiodide ions can react with the injected electrons and increase the dark current. The concentration of iodide/triiodide must be optimized.

According to physical states, compositions, and formation mechanisms, the electrolytes used in DSSCs can be broadly classified into three categories: liquid electrolytes, quasisolid electrolytes, and solid-state conductors.

4.8 TRANSPORT MECHANISM OF ELECTROLYTES IN SOLAR CELLS (DSSC)

In the electrochemical circuit of DSSCs, the electrons are transported through TiO_2 crystalline film and the holes are transported through the electrolytes or hole conductors. In this sense, electrolytes or hold conductors are hold-transport mediators. The basic function of electrolytes or hold conductors is the registration of dye as well as itself in DSSCs. For typical I^-/I_3^- redox electrolytes, the regeneration of dye can be expressed as follows

$$D^+ + I^- \xrightarrow{\text{Dye regeneration}} D + I_3^- \tag{4.1}$$

In fact, the reaction contains a series of successive reactions on the TiO_2 interface [13].

$$D^+ + I^- \rightarrow [D\ldots I] \xrightarrow{I^-} D + I_2^- \xrightarrow{I_2^-} D + I_3^- + I^- \tag{4.2}$$

$$D^+ + I_2^- \rightarrow [I_2\ldots D] \xrightarrow{I^-} D + I_3^- \tag{4.3}$$

The regenerative cycle of electrolytes is completed by the conversion of I_3^- to I^- ions on the counter electrode. The counter electrode has much catalytic activity to ensure rapid reaction and low overpotential. The reduction of I_3^- may be expressed as a successive and rapid one-electron reaction on the counter electrode.

$$I_3^- + e^- \rightarrow I^- + I_2^{-\cdot} + e^- \rightarrow 3I^- \tag{4.4}$$

The implementation of the above reactions depends on the transport of the redox mediator between the photoanode and the cathode in DSSCs.

Conductivity and diffusivity of the electrolyte are two important parameters reflecting the transport ability.

4.9 ELECTROLYTE MATERIAL USED IN SOLAR CELLS

4.9.1 Liquid Electrolytes

Liquid electrolytes possess some important features such as easy preparation, high conductivity, low viscosity, and good interfacial wetting between electrolytes and electrodes, thus a high conversion efficiency for DSSCs. Today, liquid electrolytes are still the most widely utilized transport medium for DSSCs and have produced the highest efficiency of 13% for traditional DSSCs. In DSSCs, liquid electrolytes should be chemically and physically stable, have a low viscosity to minimize charge carrier transport resistance, and be a good solvent for redox couple components. In general, a liquid electrolyte consists of three main components; solvent, ionic conductor, and additives.

4.9.2 Organic Solvents

The organic solvent is a basic component in liquid electrolytes, and it provides an environment for the dissolution and diffusion of the ionic conductor. The following general requirements should be fulfilled for the solvents in DSSCs:

(a) A melting point below −20°C and boiling point above 100°C so that the electrolyte prepared with these solvents will not evaporate under cell operating conditions.
(b) Sufficient chemical stability in the dark and under illumination. The solvent should have a wide electrochemical window so that electrolyte degradation would not occur within the range of the working potentials of both the cathode and anode.
(c) High dielectric constants so that electrolyte salts are sufficiently soluble and exist in a fully dissociated state.
(d) Low viscosity so that redox mediators possess high diffusion coefficients and the liquid electrolytes have high conductivity.
(e) Inertness with respect to the surface-attached dye, including the dye-metal oxide (or other semiconductor) bond.
(f) Poor solubility to the sealant materials as well as a low toxicity and low cost.

Hundreds of chemical compounds have been tried, and those that meet most of the requirements above fall into two categories of solvents: polar organic solvents and ionic liquids. It is notable that no single solvent can simultaneously fulfill all the requirements aforementioned; those requirements for one solvent are often, in several respects, contradictory. Therefore, when one wishes to obtain optimal DSSC performance, mixed solvents are often used. For example, a mixed solvent of acetonitrile and valeronitrile is popular, and the mixed volume ratio is either 50:50 or 85:15. The choice of solvents depends on the particular use of the DSSC under consideration.

4.9.3 Ionic Liquids

Ionic liquid (IL) is a salt in a liquid state; ionic liquids or molten salts are generally defined as liquid electrolytes composed entirely of ions. In more detail, the melting point criterion was proposed to distinguish between the molten salts with a high melting point and high viscosity and the ionic liquids with a low melting point below 100°C and relatively low viscosity. The latter are free-flowing liquids at room temperature; thus, they often are called room-temperature ionic liquids. Ionic liquids have been widely used in the electrolytes in DSSCs, due to their unique features. Those features include good chemical and thermal stability, tunable viscosity, relative nonflammability, high ionic conductivity, and a broad electrochemical potential window as well as extremely low vapor pressure, therefore providing less evaporation and leakage. Ionic liquids have two applications in electrolytes in DSSCs: one acting as solvents in liquid electrolytes and the other functioning as organic salts in quasi-solid-state electrolytes.

4.9.4 Quasisolid-State Electrolytes

Using liquid electrolytes as charge carrier transporters, the DSSC achieves great development. However, the use of liquid electrolytes causes some practical problems, such as leakage and volatilization of the solvent, photodegradation and desorption of the dye, corrosion of the counter electrode, and ineffective sealing of the cells for long-term applications. One of the methods to solve these problems is using quasisolid-state electrolytes. Quasisolid state or semisolid state is a special state of a substance between the solid and liquid states. There are three methods often used for preparing quasisolid electrolytes: (a) Liquid electrolytes are solidified by organic polymer gelators to form thermoplastic polymer electrolytes or thermosetting polymer electrolytes; (b) liquid electrolytes are solidified by inorganic gelators, such as SiO_2 or nanoclay powder, to form composite polymer electrolytes; and (c) ionic liquid electrolytes are solidified by organic polymer or inorganic gelators to form quasisolid ionic liquid electrolytes. Or, put more simply, place the quasisolid state electrolytes into four main kinds: thermoplastic polymer electrolytes, thermosetting polymer electrolytes, composite polymer electrolytes, and ionic liquid electrolytes.

4.9.5 Thermoplastic Polymer Electrolytes

In general, a polymer gel electrolyte consists of polymer or oligomer, organic solvent, and inorganic salts. Sometimes additives are added in little quantities. The main function of a polymer or oligomer is to act as a matrix or framework to gel, solidify, absorb, swell, hold, and interact with the liquid electrolyte; the polymer often is called a gelator or adsorber. The solvent, often termed a plasticizer, provides the room and surroundings for ionic salt migration. It reduces the crystallization and glass temperature of electrolytes because it exists between the adjacent polymer chains, decreases the polymer-polymer chain interaction, and increases the free volume and segmental mobility of the system. When mixing the polymer matrix with liquid electrolyte, the system gradually converts from a dilute heterogeneous system to a viscous homogeneous system or from a sol state to a gel state. In this gelation process, owing to weak interaction between the polymer matrixes (gelator) and the solvents (plastizer), a polymer gel electrolyte is obtained by the gelation, adsorption, inflation, and "entanglement network" of the polymer in a liquid electrolyte. The weak interaction included in hydrogen bonds, van der Waals, electrostatic interaction, etc., is a physical cross-linking; it depends on the temperature. Therefore, the state of this kind of electrolyte can be reversibly changed from the sol state to the gel state by controlling temperature. According to this feature, we name this kind of electrolyte "thermoplastic polymer electrolyte" (TPPE).

4.9.6 Thermosetting Polymer Electrolytes

Another kind of polymer gel electrolyte is thermosetting polymer electrolyte, in which the electrolyte is obtained by an organic molecule chemical or covalent cross-linking. This leads to the formation of a three-dimensional polymer network, and hence wrapping liquid electrolytes inside. Because the states of this kind of polymer gel electrolyte cannot reversibly change with temperature, we call this kind of electrolyte the "thermosetting polymer electrolyte" (TSPE). In appearance, the thermosetting polymer electrolytes are the same as the solid-state electrolytes. However, because some liquid electrolytes still remain in the system, we classify them as quasisolid-state electrolytes. The main difference between TPPE and TSPE lies in the fact that the former is physically cross-linking. Although the TSPEs have lower ionic conductivity than liquid electrolytes and TPPEs, the physical, chemical, and thermal stabilities of TSPEs are better than both electrolytes. So, TSPE is also an optional electrolyte for high photovoltaic performance and good long-term stability for DSSCs.

4.9.7 Composite Polymer Electrolytes

Inorganic materials such as TiO_2, SiO_2, ZnO, Al_2O_3, carbon, etc., as gelators are introduced into liquid polymer electrolytes to form quasisolid electrolytes. These quasisolid electrolytes are labeled composite polymer electrolytes. In DSSCs, the main objective for incorporating inorganic nanoparticles aims at the enhancement of long-term stability and ionic conductivity because the inorganic nanoparticles can solidify the liquid electrolyte and convert the electrolyte from a liquid state to a quasisolid state, thus enhancing the long-term stability of the electrolyte. Meanwhile, an organic and/or organic-inorganic network is constructed by the incorporation of inorganic nanoparticles in the electrolyte. Also, I^-/I_3^- ions are able to align and transport on the inorganic particles network, leading to an acceleration of the charge-transport dynamics.

4.9.8 Blend Biopolymer Electrolytes (BBPE) for DSSC

Rudhziah et al. [14] prepared blended biopolymer electrolytes based on kappa-carrageenan and cellulose derivatives for a dye-sensitized solar cell. In this work, carboxymethyl kappa-carrageenan (CMKC)/carboxymethyl cellulose (CMCE) were blended for use as electrolytes in a solid-state DSSC application. They have synthesized CMKC and CMCE from k-carrageenan and kenafiber. The synthesized CMKC and CMCE were then blended and doped with ammonium iodide (NH_4I). NH_4I was chosen as a doping salt because it has the potential to be used as a salt in the polymer electrolyte together with iodine (I_2), especially for the DSSC electrolytes. It also has low lattice energy and a larger anionic size, which are favorable for providing high ionic conductivity.

4.10 FABRICATION OF SOLID-STATE DSSC DEVICE

A solid-state, dye-sensitized solar cell with an active area about 1 cm^2 was fabricated as follows. In order to prepare the DSSC photoelectrode, the TiO_2 paste was spread on fluorine-doped tin oxide (FTO) conducting glass using a doctor blade; this was followed by a sintering process at 450°C for 30 min. This electrode was then immersed in the solution of the dye N719 for 24 h. Meanwhile, the platinum FTO glass counter electrode was prepared by a brush painting technique and heating at 450°C for 30 min. For redox couple formation in DSSC, the prepared electrolyte solution was then doped with 0.02 M of I_2. This electrolyte solution was cast onto the TiO_2/dye photoelectrode with 0.02 M of I_2. This electrolyte solution was cast onto the TiO_2/dye photoelectrode and heated at 50°C to form film. Next, the TiO_2/dye photoelectrode with electrolyte film was then assembled with a platinum counter electrode. The photocurrent density-photovoltage characteristics of the solid-state DSSC were obtained under a solar simulator.

4.10.1 Polymer-Salt Interaction and its Conductivity Studies

A polymer-salt interaction study proves from FTIR. From FTIR they have confirmed the interactions between NH_4^+ ions or H^+ ions and oxygen atoms of CMKC/CMCE chains. The temperature dependence of ionic conductivity of the CMKC/CMCE-NH_4I was investigated in the different temperature. The conductivity of the system is observed to increase gradually with an increase in temperature. This illustrates that the ionic conduction in the biopolymer electrolyte follows the Vogel-Tammann-Fulcher (VTF) theory. It states that the cooperative process in the ionic conduction mechanism where the ionic hopping mechanism is coupled with higher polymer segmental mobility in an amorphous phase. The rise in temperature enhances the flexibility of polymer chains as a result of an increase in the thermal oscillation mode. The movement of polymer segments is thus improved. Fast segmental mobility could promote the breaking down of the weak interaction between the proton and the polar group in the polymer. The detachment of the proton creates more free voids for the ions to jump from an interstitial site to another adjacent equivalent site and eventually generate ionic transportation [15].

The ionic transference number for the electrolytes containing 20 and 30 wt.% NH_4I salt has been determined by the DC polarization method. The values of the ionic transference number of the electrolytes containing 20 and 30 wt.% salt are found to be 0.9933 and 0.9995, respectively. This suggests that the charge transport in the electrolytes was primarily ionic and only a negligible contribution came from the electron. From the photocurrent-photo potential they calculated that the energy conversion efficiency (η) obtained is 0.13%. The value is low and may be due to poor contact at the electrode-electrolyte interface. To enhance the efficiency of DSSC, the electrolyte should be prepared in a gel form. It is believed that an electrolyte in gel form will make better electrode-electrolyte contact. Finally, they have proved that blended derivatives of biopolymer-based k-carrageenan and cellulose have been successfully synthesized and used for DSSC. The fabricated DSSC using this electrolyte showed a response under light intensity of 100 mV cm^2 with an efficiency of 0.13%. The results suggested that this blend biopolymer system has favorable properties for DSSC applications.

4.10.2 Composite Biopolymer Electrolytes (CBPE) for DSSC

Willgert et al. describe the preparation of solvent-free nanocomposite gel electrolytes in combination with copper (I)-based, dye-sensitized solar cells (DSSCs). The electrolytes comprise poly(ethylene oxide) (PE) and cellulose nanocrystals (CNCs) and an I_3^-/I^- redox shuttle. They have proved DSSCs containing the nanocomposites and the copper (I)-based dye show robust stability over time and after 60 days.

4.10.2.1 Preparation of the Nanocomposite Electrolytes

An oligomer solution was prepared in a particular ratio and was shaken until it completely dissolved. To this carbon nanotube, water suspension was added at desired concentrations. The samples were then placed in an oven at 70°C overnight, followed by UV irradiation for 30 min. After polymerization, a liquid I_3^-/I^- electrolyte solution was added to each sample, which was allowed to swell for 1 h. Finally, each sample was dried under high vacuum to remove all acetonitrile.

4.10.2.2 Fabrication of DSSC Based on Composite Electrolytes

TiO_2 electrodes were carefully washed with EtOH and subsequently heated to 455°C for 20 min. After cooling to 80°C, the electrodes were immersed in the phosphoric acid anchoring ligand solution for 24 h. The ligand-immobilized electrodes were then carefully washed with DMSO, followed by a CH_3CN solution. After this time, the dye-sensitized electrodes were taken out of the solutions, rinsed thoroughly with CH_3CN, and dried under an N_2 stream. Then 5 mg of the composite electrolyte paste was put onto the electrodes. Pt-coated cathode electrodes that had been cleaned the same way as the anodes were sealed together with the anode using a hot melt sealing foil with an initial thickness of 60 μm.

4.10.2.3 DSSC Performances Based on Composite Electrolytes

The I_3^-/I^- electrolyte/nanocomposite paste was combined in DSSCs with the copper (I) dye. The most fundamental DSSC characterization is by *J-V* measurements to determine the photo conversion efficiency (η), the fill factor (FF), the open circuit potential (V_{oc}), and the short circuit current density (J_{sc}) of the DSSCs. The amount of cellulose nanocrystals added has a dramatic effect on efficiency. The efficiency increases 19-fold when going from the neat PEO sample to the sample containing 80% cellulose nanocrystal; this is the best performing nanocomposite device in the composite biopolymer electrolytes. One of the central parameters of this study is the stability of the DSSC over time, where the nanocomposite samples show outstanding durability with respect to a reference DSSC containing liquid electrolyte as well as compatibility with copper-based dyes. After 60 days, the DSSC shows 80% efficiency compared to initial days. A series of DSSCs sensitized with the bis(diamine)copper(I) dye with a nanocellulose composite incorporated into the I_3^-/I^- and the amount of cellulose nanocrystal added has a large impact on the performances of the cells, with a larger amount leading to a significant improvement in photo-conversion efficiency. Very few reports are available on composite electrolytes for DSSC.

4.10.3 GEL Biopolymer Electrolytes (GBPE) for DSSC

Gong et al. [16] explain the gel polymer electrolyte containing I^-/I_3^- redox couple was prepared using polyethylene glycol (PEG) as the polymer matrix and propylene carbonate (PC) as the organic solvent by the sol-gel method. Potassium iodide (3.25 g) and iodine (0.5 g) were dissolved with 30 mL PC in a 250 mL four-neck flask. The mixture was stirred for about 30 min at room temperature, and 25.5 g PEG-20000 was added into the flask. In order to ensure the chemical reaction happened in the inert atmosphere, nitrogen (N_2) gas was provided as ambience. Reactions occurred at 100°C for 24 h to form a homogeneous colloidal liquid. At the end of the reaction process, the colloidal liquid was cooled to 60°C to form a polymer gel electrolyte. The prepared electrolyte was heated to melt and was slowly added dropwise to the TiO_2 surface so that the electrolyte could thoroughly penetrate into the porous TiO_2 film. A sandwiched cell structure was finally assembled by pressing the Pt electrode on the top. The fabricated solar cells were then sealed with a low cost, commercially available epoxy adhesive. They have proved the gel polymer electrolyte gives an open-circuit voltage of 0.7 V and a short-circuit current of 8.1 mA cm^{-2} at an incident light intensity of 100 mW cm^{-2}.

Federico Bell et al. [17] reported on the synthesis and characterization of a new acrylic/methacrylic-based gel-polymer electrolyte membrane for dye-sensitized solar cells. However, it was then found to have a poor light-to-electricity conversion efficiency of 3%. A bio-based chitosan/PVdF-HFP gel polymer electrolyte was also prepared by Yahya et al. [18]. The quasisolid-state electrolyte tested in DSSCs with an optimized weight ratio of PVDFHFP: chitosan (6:1) with an ionic liquid electrolyte 1-methyl-3-propylimidazolium iodide ($PMII/KI/I_2$) has shown the highest power conversion efficiency of 1.23% under 100 mW cm^{-2} light illumination with ionic conductivity of 5.3×10^{-4} S cm^{-1}. This demonstrates the potential for using sustainable bio-based chitosan polymers in DSSC applications.

Khanmirzaei et al. [19] prepared hydroxypropyl cellulose (HPC)-based non-volatile gel polymer electrolytes for dye-sensitized solar cells using ionic liquid.

The highest ionic conductivity of 7.37×10^{-3} S cm^{-1} is achieved introducing 100% of 1-methyl-3-propylimidazolium iodide (MPII) as the ionic liquid (IL), with respect to the weight of HPC. The gel polymer electrolyte with 100 wt.% of the MPII ionic liquid shows the best performance and energy conversion efficiency of 5.79%, with short-circuit current density, open-circuit voltage, and fill factor of 13.73 mA cm^{-2}, 610 mV, and 69.1%, respectively.

4.10.4 Solid Bopolymer Electrolytes (SBPE) for DSSC

Solid biopolymer electrolytes used in electrochemical devices, normally polysaccharide, are the most prominent biopolymers used for developing efficient electrochemical devices. In this chapter we will discuss widely used polysaccharides for DSSC applications.

4.10.4.1 Agarose/Agar

Agarose is a linear polymer consisting of alternating beta-D-galactose and $1 \rightarrow 4$-linked 3,6-anhydro-alpha-L-galactose units. It has very few sulfate groups. The gelling temperature ranges from 32°C to 45°C, and the melting temperature range is normally 80–95°C. Methylation, alkylation, and hydroxyalkylation of the polymer chain can change the melting and gelling temperatures [20]. Generally it is insoluble in cold water, but it swells considerably. However, it can easily dissolve in hot water (H_2O). It also dissolves in solvents such as DMF (dimethylformamide), DMSO (dimethylsulfoxide), dimethylacetamide (DMAc), glycol, orthophosphoric acid, *N*-Methyl-2-pyrrolidone (NMP), etc in temperatures between 95°C and 100°C. The maximum conductivity reported in the literature is around 10^{-2} S cm^{-1}.

4.10.4.2 Carrageenan

Carrageenan is obtained from the red seaweed of the class rhodophyceae. It is a group of linear galactan with an ester sulfate content of 15%–40% (*w*/w) and containing alternating $(1 \rightarrow 3)$-α-D and $(1 \rightarrow 4)$-β-D-galactopyranosyl linkages. The three types of commercially available carrageenan are *k*, *l*, and *λ*. Anionic polysaccharides with molecular weight between 100,000 and 1,000,000 form gels with potassium or calcium ions. Carrageenan can also easily dissolve in hot water even at room temperature. The ionic conductivity of carrageenan was found in the range of 10^{-7}–10^{-3} S cm^{-1} in water and DMSO.

4.10.4.3 Alginate

Alginate is obtained from the brown seaweed of the class Phaeophyceae as a structural material. Linear polysaccharide is composed of β-D-mannuronopyransyl and α-L-guluronopyranosyl units. The units occur in M blocks (containing mannuronopyranose residues), G blocks (containing guluronopyranose residues), or MG blocks. The ratio of G-, M-, and MG-blocks affects the gel strength, calcium reactivity, and other properties. Alginate forms gels with calcium ions. It dissolves slowly in water and forms a viscous solution, but is insoluble in ethanol and ether. Gel polymer electrolytes (GPEs) based on sodium alginate plasticized with glycerol showed ionic conductivity 3.1×10^{-4} S cm^{-1} for the samples of $LiClO_4$ and 8.7×10^{-5} S cm^{-1} for the samples with CH_3COOH at room temperature.

4.10.4.4 Pectin

Pectin is found in all land-based plants as a structural material. Commercial pectin is extracted from citrus peels, apple pomace, sugar, beets, or sunflower heads. A linear chain of galacturonic acid units has a molecular weight of about 110,000–150,000. Pectin is soluble in pure water. Monovalent cation salts of pectinic and pectic acids are usually soluble in water; di- and trivalent cations salts are weakly soluble or insoluble. The plasticized pectin and $LiClO_4$-based gel electrolytes were prepared and analyzed by spectroscopic, thermal, structural, and microscopic analyses. The best ionic conductivity value of 2.53×10^{-2} S cm^{-1} was obtained at room temperature.

4.10.4.5 Cellulose

Cellulose is the most abundant polymer available worldwide. Cellulose is composed of polymer chains consisting of unbranched β $(1 \rightarrow 4)$ linked D-glucopyranosyl units (anhydroglucose unit). Nowadays, there are various procedures for the extraction of cellulose microfibrils such as pulping methods, acid hydrolysis, steam explosion, etc. The samples of HEC (hydroxyethylcellulose) plasticized with glycerol and lithium trifluoromethane sulfonate ($LiCF_3SO_3$) salt showed ionic conductivity in the range of 10^{-4}–10^{-5} S cm^{-1}. The best ionic conductivity obtained is 4.9×10^{-3} S cm^{-1} at room temperature by using lithium perchlorate ($LiClO_4$)-doped biopolymer cellulose acetate (CA), reported by Selvakumar et al. [21]

4.10.4.6 Plant Seeds, Plant Tubers, Root and Cereal Starch

The principal crops used for starch production include potatoes, corn, and rice. In all these plants, starch is produced in the form of granules, which vary in size and somewhat in composition from plant to plant. The starch granule is essentially composed of two main polysaccharides: amylose and amylopectin with some minor components such as lipids and proteins. Amylose is a linear molecule of (1→4)-linked α-D-glucopyranosyl units and molecular weights ranging from 105 to 106 g mol^{-1}. Amylopectin is a highly branched molecule composed of chains of α-D-glucopyranosyl residues linked together mainly by (1→4)-linkages but with (1→6) linkages at the branch points and having molecular weights ranging from 106 to 108 g mol^{-1}. Amylose is water soluble but Amylopectin is insoluble in cold water and swells in it, thereby giving rise to a thick paste upon boiling with water. It is a biopolymer containing about 23% starch (20%–25% amylose and 75%–80% amylopectin). It is clear that adding KI in an arrowroot matrix enhances the ionic conductivity. Conductivity maxima were obtained by doping the NaI and KI concentration where conductivity values approach at 6.7×10^{-4} and 1.04×10^{-4} S cm^{-1}, respectively. Adding (Glycerol+LiCl) and KI in the Sago Palm matrix enhances the ionic conductivity and conductivity maxima was obtained by adding LiCl where the conductivity value approached to 10^{-3} S cm^{-1} with LiCl and 3.4×10^{-4} S cm^{-1} for KI [22,23].

Corn starch-based biopolymer electrolytes have been prepared by the solution casting technique. Lithium hexafluorophosphate ($LiPF_6$) and 1-butyl-3-methylimidazolium trifluoromethanesulfonate (BmImTf) were used as lithium salt and ionic liquid, respectively. In another study, ionic liquid, 1-butyl-3-methylimidazolium hexafluorophosphate (BmImPF6) was doped into the corn-based biopolymer matrix; a maximum ionic conductivity of 1.47×10^{-4} S cm^{-1} was reported for this system. Blend biopolymer electrolyte was prepared using starch/chitosan with different ratio and varying lithium perchlorate ($LiClO_4$). The highest ionic conductivity achieved at room temperature was 3.7×10^{-4} S cm^{-1} [24].

Rice starch doped with LiI was prepared using a solution casting method; at room temperature the highest ionic conductivity achieved was around 4.7×10^{-5} S cm^{-1}. Hence, they can interact with protons or lithium ions leading to ionic conduction. Among different natural polymers, the starch-based biopolymer electrolyte presented good optoelectrochemical characteristics and can be applied to electrochemical devices. The ionic conductivity results obtained for these biopolymer electrolytes varied from 10^{-6} to 10^{-4} S cm^{-1} at room temperature, Conductivity reaches 8.1×10^{-3} S cm^{-1} for cassava doped with lithium perchlorate. The amount of acetic acid and NH_4NO_3 was found to influence the proton conduction. Wheat can easily dissolve in acetic acid and room temperature conductivity was found in the order of 10^{-5}–10^{-4} S cm^{-1}.

4.10.4.7 Chitin and Chitosan

Chitosan is a linear polysaccharide consisting of β (1 → 4) linked D-glucosamine with randomly located *N*-acetylglucosamine groups depending upon the degree of deacetylation of the polymer. Chitin is basically found in the shells of crabs, lobsters, shrimp, and insects. Chitosan is the deacylated derivative of chitin. Chitin is insoluble in its native form but chitosan is water-soluble. Chitosan is soluble in weakly acidic solutions, resulting in the formation of a cationic polymer with a high charge density. Therefore, it can form polyelectrolyte complexes with a wide range of anionic polymers. Chemical modification of chitosan can significantly affect its solubility and degradation rate. The electrical properties of polymer electrolytes based on chitosan doped with lithium and ammonium salts have been reported in literature. Conductivities of the order of 5.5×10^{-3} S cm^{-1} at room temperature were reported for chitosan doped with $LiClO_4$ salt [25]. $LiMn_2O_4$-doped, biopolymer-based chitosan with carbon has reported that a biopolymer-in-salt based electrolyte achieves the best ionic conductivity: 3.9×10^{-3} S cm^{-1} at room temperature [26].

4.10.4.8 Gum Arabic

A gummy exudate obtained from Acacia trees with a molecular weight of about 250,000 is highly soluble with low viscosity, even at 40% concentration. Gum arabic, for example, *Acacia arabica* and Acacia babul, exhibits a conductivity of $\sim 1.5 \times 10^{-6}$ S cm^{-1} after drying. Gum Arabica produces salt complexes with inorganic materials such as $FeSO_4$, [K_2SO_4, $Al_2(SO_4)_3$, H_2O], $LiClO_4$, iodine, etc., and functions as a proton conductor through hydronium ions H_3O^+.

4.10.4.9 Gum Tragacanth

Gum Tragacanth is an exudate of Astragalus, a perennial short brush in Asia. It is slightly acidic and found as Ca, Mg, or Na salts. It contains neutral highly branched arabinogalactan and tragacanthic acid (linear (1 → 4)-linked α-D-galacturonopyranosyl units, with some substitutions). It is highly viscous with some emulsification properties. The highest conductivity reported for an NaOH-based biopolymer is 88.8×10^{-3} S cm^{-1} at room temperature. Xanthan gum is prepared through culturing Xanthomonas campestris, a single-cell organism producing gum as a protective coating. A trisaccharide side chain is attached to alternate D-glucosyl units at the *O*-3 position. The side chain consists of a D-glucuronosyl unit between two D-mannosyl units. Molecular weight is about 2,000,000–3,000,000. Its viscosity is stable at a wide temperature and pH range. Among the systems studied, that is, Gum Xanthan + PVP, Gum tragacanth + PVP, and Gum Acacia+ PVP, Gum Acacia + PVP presents better compatibility as it has a stronger intermolecular interaction. In the same manner, among Gum Xanthan + PEG, Gum Acacia + PEG, and Gum tragacanth + PEG systems, Gum Tragacanth + PEG has better compatibility.

4.10.4.10 Gellan Gum

Gellan gum is prepared by culturing Pseudomonas elodea. It is composed of a four-sugar repeating sequence containing one D-glucuronopyranosyl, two D-glucopyranosyl, and one L-rhamnopyranosyl unit. Its molecular weight is about 1,000,000–2,000,000. It requires either monovalent or divalent cations to form a gel. The ionic conductivity measurements revealed that the ionic conductivity of the Gellan gum doped with 40 wt.% of (lithium trifluoromethanesulfonate) $LiCF_3SO_3$ electrolyte varies with the salt concentration, reaching the highest conductivity value of 5.4×10^{-4} S cm^{-1} at room temperature. When doped with LiI, it exhibits ionic conductivity of 3.8×10^{-4} S cm^{-1} at room temperature. To achieve good ionic conductivity, plasticizers such as glycerol, ethylene glycol, ethylene carbonate, and propylene carbonate are used; adding lithium salts of $LiClO_4$, $LiBF_4$, $LiCF_3SO_3$, LiI/I_2, or acetic acid promoted the proton conduction. The gellan gum (GeG) was prepared and it reached maximum ionic conductivities such as 5.1×10^{-3} S cm^{-1} while activated energies were between 0.14 and 0.19 meV at different temperatures [27].

4.10.4.11 Carboxymethyl Cellulose (CMC)

Carboxymethyl cellulose (CMC) is prepared by soaking cellulose in aqueous sodium hydroxide and having it react with monochloroacetic acid. Carboxymethyl cellulose doped with lithium perchlorate and plasticizer polycarboxylate-based transparent solution of $CMC/LiClO_4/PC$ is reported and ionic conductivity of the biopolymer electrolyte is found to be 2×10^4 S cm^{-1}. In another system, an oleic acid-based biopolymer electrolyte and an NH_4Br-based electrolyte achieved ionic conductivity values of 2.1×10^{-5} and 1.1×10^{-4} S cm^{-1}, respectively. CMC is doped with a different concentration of DTAB/EC via the solution casting technique. The highest ionic conductivity was found to be 4.1×10^{-3} S cm^{-1} at room temperature [28].

4.10.5 Introduction to Battery

Passage of electricity through matter may be associated with magnetic effects, heating effects, or chemical reactions. Materials that conduct electricity may be classified into two types: metallic or electronic and electrolytic. In a metallic conductor, for example, copper wire, the electricity is carried by the electrons; there is no appreciable movement of matter or a chemical reaction as a result of the passage of electricity. On the contrary, in an electrolytic conductor (otherwise called an electrolyte), the current is carried by ions. The passage of electricity through an electrolyte always results in chemical reactions and transfer of matter.

4.10.6 E.M.F. and Resistance

The electrical driving force responsible for the flow of current is called the electromotive force (e.m.f.). Between any two points in the circuit carrying the current, there exists a potential difference. The algebraic sum of all the potential differences is the total e.m.f.

Every bit of matter is endowed with the ability to hinder the flow of electricity to a different extent. This hindrance to the flow of electricity is called resistance.

The magnitude of current flowing through a system is given by Ohm's law, which states that the current strength (I) is directly proportional to the e.m.f. (E) and inversely proportional to the resistance (R), that is,

$$I = \frac{E}{R}$$

The SI units of current, e.m.f., and resistance are ampere, volt, and ohm, respectively.

4.10.7 Quantity of Electricity and Electrical Energy

Quantity of electricity (q) is given by the product of current strength and time:

$$q = It$$

One ampere of current flows through a system for 1 s with a certain quantity of electricity. The passage of electricity through some conductors is accompanied by the liberation of heat and has been found to be proportional to the quantity of electricity passed and the potential difference. Heat liberated may be equated to electrical energy (Q):

$$Q = EIt = I^2RT$$

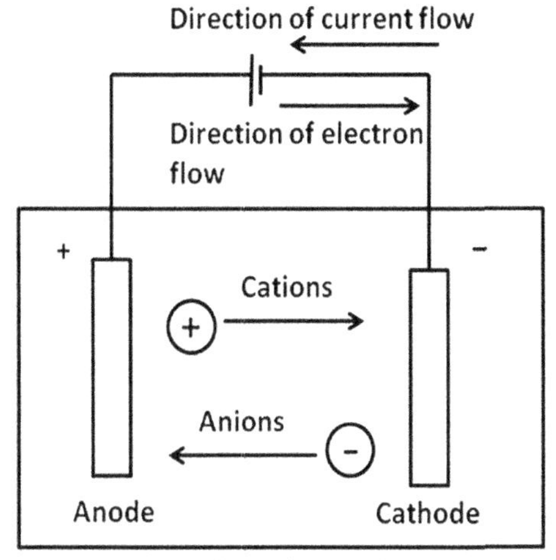

FIG. 4.2 Electrolytic cell.

4.10.8 Electrolytic Conduction

In order to pass a current of electricity through an electrolyte, two electrodes-generally two plates of metals or carbon-are inserted into the electrolyte and the electrodes are then connected to a source of e.m.f. Positive current enters the solution at the anode and leaves from the cathode. The cations move through the solution toward the cathode while the anions move toward the anode (Fig. 4.2).

The function of the applied e.m.f. is to direct the ions to the appropriate electrodes. The electrons move from the anode to the cathode outside the cell. The anions supply the electrons at the anode; the same number of electrons is removed at the cathode by the cations. The latter two processes represent the chemical reaction that takes place at the electrodes due to the passage of electricity.

4.10.9 Faraday's Laws of Electrolysis

In electrolysis, the chemical change may involve the evolution of a gas at the anode or cathode or the deposition of a metal at the cathode.

4.10.9.1 Faraday's First Law

The amount of chemical change produced by an electric current, that is, the amount of any substance deposited or dissolved, is proportional to the quantity of electricity passed.

4.10.9.2 *Faraday's Second Law*

The amounts of different substances deposited or dissolved by the same quantity are proportional to their chemical equivalent weights. Combining these two laws,

$$w = Ite/F$$

where w is the weight in grams of the material deposited or dissolved at an electrode, e is the equivalent weight of the material, and $1/F$ is proportionality constant.

The significance of the constant F may be understood by considering the passage of a quantity of electricity (It) equal to F through a solution; then $w = e$. In other words F represents the quantity of electricity required to deposit or dissolve a 1 g equivalent of any substance.

4.10.9.3 *Significance of Faraday* (F)

Because one faraday of electricity is required to discharge a 1-g equivalent of any substance, it must be equal to the total charge on one equivalent of ions, the charge on one Avogadro number of univalent ions. Therefore,

$$\begin{aligned} F &= N \times \text{electronic charge} \\ &= 6.023 \times 10^{23} \times 1.602 \times 10^{-19}\text{C} = 96488\,\text{C} \end{aligned}$$

4.10.10 Electrode Potential

Consider $Zn(s)/ZnSO_4$

$$\text{Anodic process} : Zn(s) \rightarrow Zn^{2+}(aq)$$

$$\text{Cathodic process} : Zn^{2+}(aq) \rightarrow Zn(s)$$

$$\text{At equilibrium} : Zn(s) \leftrightarrow Zn^{2+}(aq)$$

Metal has a net negative charge and a solution has an equal positive charge, leading to the formation of a Helmholtz electrical layer. The electric layer on the metal has a potential Ø (M). The electric layer on the solution has a potential Ø (aq). The electric potential difference between the electric double layers is generated across the electrode/electrolyte interface of a single electrode or half cell. When a solid is in contact with a liquid, a double layer of ions appears at the surface of separation. The double layer consists of two parts: one is fixed and the other is a diffuse layer.

- Fixed layer: One part of the double layer is fixed on the surface of the solid (it consists of either positive ions or negative ions).
- Diffuse or mobile layer: This layer consists of ions of both signs but its net charge is equal and opposite to that on the fixed part of the double layer.

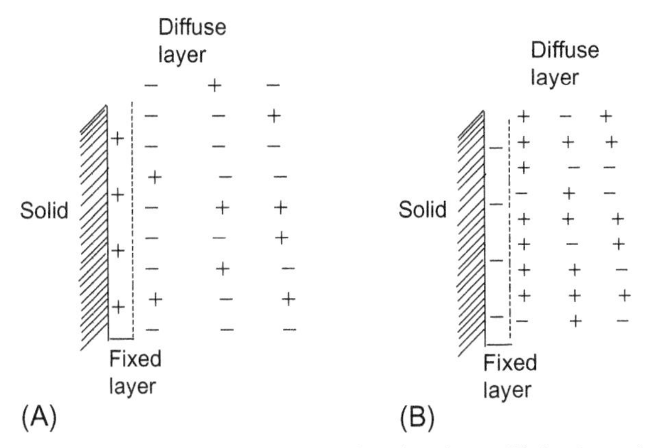

FIG. 4.3 Helmholtz electrical layer: (A) fixed positive ions layer, (B) fixed negative ions layer.

From Fig. 4.3, we see that the net charge on the diffuse layer is equal and opposite to that on the fixed layer.

4.10.10.1 Electrode Potential

This is the potential developed at the interface between the metal and the solution when it is in contact with a solution of its ions.

4.10.10.2 Single Electrode Potential

Whenever a metal is in contact with its own ions, it has a natural tendency to lose or gain electrons. This tendency leads to a gain of potential difference across the solution and electrode; this is known as the single-electrode potential.

4.10.11 Basic Term of Batteries

- **Cell**: Single arrangement of two electrodes and an electrolyte to provide electricity due to a redox reaction.
- **Battery**: Combination of two or more cells arranged in series or parallel to produce more voltage.
- **Charging:** Process in which a battery is restored to its original charged condition by the reversal of current flow.
- **Discharging**: Process by which a battery delivers electrical energy to an external load.
- **Separator**: Physical barrier between anodes and cathodes in a battery to prevent internal short circuiting. Separators must be ionically conducting but electronically insulating and inert in the battery environment.
- **Voltage:** Voltage can be defined as the amount of pressure of electrons that passes from a negative electrode to a positive electrode. Mathematically,

$$V = IR$$

where V = voltage; I = current; R = resistance

- **Current**: Current is a measure of the rate at which the battery is discharging.
- **Capacity:** The capacity is the charge or the amount of electricity that may be obtained from the battery and is given in ampere hours (Ah).

$$\text{Capacity} = I \times t$$

where I = current; t = time

- **Electricity storage density**: Electrical storage density is the amount of electricity per unit weight that the storer can hold; it's the capacity per unit weight of the battery.
- **Power (Watts)**: Power is the power per unit weight of the battery.

Power (Watts) = EI, where E = voltage, I = current.

- **Cycle life**: Primary batteries are designed for a single discharge, but a secondary battery is rechargeable. The cycle life is the number of complete charge/discharge cycles a battery can perform before its capacity fails below 80% of its initial capacity.
- **Shelf life**: The time an inactive battery can be stored before it becomes unusable, that is, the length of time a battery can remain without losing its energy capacity.
- **Design life**: Elapsed time before a battery becomes unusable whether it is in active use or inactive.

4.10.12 Need for Biopolymer Electrolytes in Batteries

As already mentioned, an efficient battery needs electrolytes with high ionic conductivity. Solid polymer electrolytes (SPEs) have been widely examined to substitute for conventional liquid electrolytes because of their attractive properties, such as the ability to eliminate problems of corrosive solvent leakage, a wide electrochemical stability range, light weight, ease of processability, and excellent thermal stability as well as low volatility. The most commonly studied polymer electrolytes for batteries are high molecular weight polyethylene oxides (PEO) complexed with Li salts. PEO qualifies as a host polymer for electrolytes because of its high solvating power for lithium salts and its compatibility with a lithium electrode. However, one of the major drawbacks of PEO-based solid polymer electrolytes is their low ionic conductivity (10^{-7}–10^{--8} S cm^{-1}) at ambient temperature, which limits their practical applications. Most SPEs are based on hydroxethyl cellulose, starch, chitosan, agar-agar, pectin, gelatin, etc., wherein the ionic conductivity has been reported to be on the order of 10^{-4} S cm^{-1} at room temperature. Cellulose constitutes are the most abundant renewable polymer source available worldwide today. It is principally used as construction material in the form of wood, a textile fiber such as cotton, or in the form of paper and board. Recently, cellulose and its derivatives have been successfully applied in rechargeable batteries for the production of electrodes or separators or as reinforcing agents in solid polymer electrolytes.

4.10.13 Performance Requirements and Ion Transfer Mechanisms

In the battery, the polymer electrolyte is sandwiched between the anode (lithium metal, carbon, etc.) and the composite cathode, acting as both electrolyte and separator. The polymer membrane plays a crucial role in the performance of lithium polymer batteries. From a practical application point of view, polymer electrolytes should possess the following properties:

(a) High ionic conductivity: It should be a good ionic conductor and electronic insulator so that ion transport can be facilitated and self-discharge can be minimized. The ionic conductivity is the determinant factor of the internal impedance and electrochemical behavior at different charge/discharge rates. The ionic conductivity of aprotic organic solvents, which contain lithium salt, can reach 10^{-2}–10^{-3} S cm^{-1}. Polymer electrolytes should possess conductivities approaching or beyond 10^{-4} S cm^{-1} at ambient temperature to quickly achieve charge/discharge.

(b) Appreciable Li^+ transference number: If possible, the Li^+ transference number should be close to unity in an electrolyte system. A large Li^+ transference number can reduce the concentration polarization of electrolytes during the charge/discharge process, thus producing high power density. Reducing the mobility of anions can largely increase the Li^+ transference number. Two approaches have been reported to effectively reduce the mobility of anions. The one is to anchor anions to the polymer backbone, and the other is the introduction of anion receptors that are selectively complexed with the anions in electrolytes.

(c) Good mechanical strength: The mechanical strength of polymer electrolytes is the most important factor that needs to be taken into account in large-scale manufacturing of the batteries. So the polymer electrolytes should not be brittle like certain ceramics, but be able to relax elastically when stress arises in the process of manufacturing, cell assembly, storage, and usage. Some feasible approaches have been introduced to strengthen the dimensional stability of the electrolyte membranes such as adding inorganic fillers, cross-linking, physical support by a polyolefin membrane, etc.

(d) A wide electrochemical stability window: An electrochemical window is defined as the difference between the potentials of oxidation reaction and reduction reaction. As an electrolyte, the primary requirement is being inert to both electrodes, which means that the oxidation potential must be higher than the embedding potential of Li^+ in the cathode and the reduction potential must be lower than that of lithium the metal in anode. Thus, the polymer electrolyte should have an electrochemical window up to 4–5 V versus Li/Li^{+} to be compatible with both electrode materials.

(e) Excellent chemical and thermal stability: The polymer electrolytes should be inert to battery components such as the anode, cathode, cell separator, current collectors, additives, and cell packaging materials. Excellent thermal stability ensures the safe use of a battery even in the case of electrical (shorting, over charge) or thermal (flame) abuse.

4.10.14 Ion Transfer Mechanism

In the past three decades, there has been much interest in the mechanism of ionic conduction in polymer electrolytes. For solid polymer electrolytes, the polymer matrix should first have the ability to dissolve/complex the lithium ions. Polymers with sequential polar groups, such as —O—, =O, —S—, —N—, —P—, —C=O, and C=N, may dissolve lithium salts and form polymer-salt complexes. Further, to facilitate the dissociation of inorganic salts in polymer hosts, the lattice energy of the salt should be relatively low and the dielectric constant of the host polymer should be relatively high. Typically, ionic conductivity is proportional to the effective number of mobile ions, the elementary electric charge, and the ion mobility. The effective number of mobile ions (free ions) depends on the degree of salt dissociation in the polymer host. It is generally acceptable that the ion transport in dry solid polymer electrolytes occurs only in amorphous regions above the glass transition temperature (T_g), with the segmental motion of chains playing a significant role in the ionic conductivity. The detailed ionic conductive mechanism can be expressed as follows: lithium ions located at suitable coordination sites (e.g., —O— in polyethylene oxide, —CN in polyacrylonitrile, —NR in polyamide, etc.) in the polarity chains of the polymer. The polymer chains undergo constant local segmental motion and result in the appearance of free volume. Lithium ions migrate from one coordination site to new sites along the polymer chains or hop from one chain to another through these free volumes under the effect of the electric field. However, this concept is tuned by Bruce et al. and they said that ionic conductivity in the static, ordered environments of the crystalline phase can be greater than that in the equivalent amorphous material above T_g. They proposed that, in the crystalline phase of $P(EO)_6$-LiX (X=PF_6, AsF_6, SbF_6), pairs of PEO chains fold to form cylindrical tunnels within which the Li-cations are located and coordinated by ether oxygen while the anions are located outside these tunnels in the interchain space and do not coordinate the cations [29].

4.10.15 Solid Biopolymer Electrolytes (SBPE) for Batteries

4.10.15.1 Fabrication of a Battery

The highest conducting film in the SBPE was used to fabricate the battery. Batteries are composed of a cathode and anode as the electrodes, a metal salt solution (solid polymer electrolyte) as the electrolyte, a separator, and functional additives that convert chemical into electrical energy. Copper and aluminium foils are used as substrates for coating anode and cathode materials in the rechargeable battery. The chemical reactions in the battery cause a build up of electrons at the anode. This results in an electrical difference between the anode and cathode. The electrolyte keeps the electrons from going straight from the anode to the cathode within the battery. The common anode materials used in batteries are titanium oxides, graphite, porous carbons, alloys (Ge, Al, Sn, Sb, etc.) metal oxide and oxy salts, pure metal foils, etc. The cathode materials used in batteries are vanadium oxide, molybdenumsulfide, molybenumoxide, manganeseoxide, silicates, $LiCoO_2$, $LiFePO_4$, $LiMn_2O_4$, etc.

4.10.16 Biopolymer Material-Based Carboxymethyl Cellulose in a Rechargeable Proton Battery

The solid polymer electrolyte has improved ionic conductivity enough for them to succeed in being applied commercially. Much research has been devoted to the preparation of solid polymer electrolytes made of various materials. One of the well-known ones is synthetic polymer materials (petroleum resources), which attracted great attention to be used in the making of proton-conducting solid polymer electrolytes such as poly(ethylene oxide), poly(acrylic acid), and poly(vinyl alcohol). However, these polymers are high in cost and the depletion of petroleum resources coupled with increasing environmental regulations are acting synergistically to provide for new materials and products that are compatible with the environment and independent of fossil fuels. For these reasons, a lot of effort has been made to develop electrolytes using natural biopolymer materials that have gained more and more attention, owing to their abundance in nature, their environmental friendliness, and their potential to substitute for petrochemicals. Therefore, the present work focuses on developing a proton-conducting biopolymer electrolyte (BPE) by using natural biopolymer materials, namely carboxymethyl cellulose (CMC), as the host polymer while NH_4Br has been chosen as the ionic dopant because ammonium salts are considered good proton donors to the biopolymer matrix.

4.10.16.1 Conduction Mechanism of CMC

In the polymer-ammonium salt system, the conducting species is H-ion, which originates from the ammonium ion. The interaction process occurs through structure diffusion (Grotthus mechanism) where the exchange of ions happens between the complexed sites. Because this biopolymer electrolyte system uses ammonium salt as the dopant agent, a lone proton migration (H^+) mechanism is more possible due to the following explanation where two of the four hydrogen atom of NH_4^+ ions are bound identically, one hydrogen atom is bound more strictly, and the fourth is bound more weakly. The weakly bound H of NH_4^+

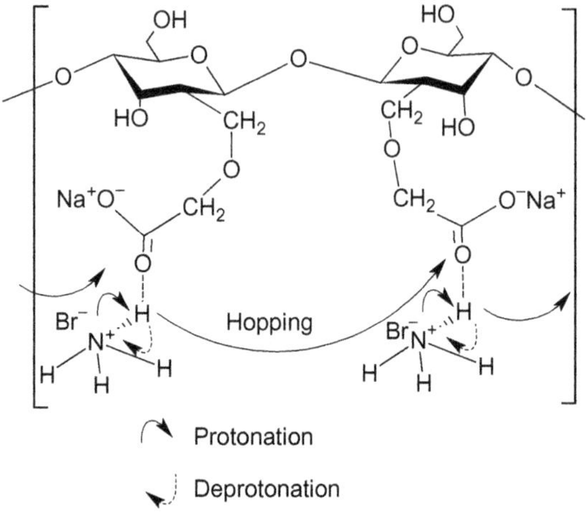

FIG. 4.4 Conduction mechanism of CMC.

can easily be dissociated and these H^+ ions can hop from one site to another, leaving a vacancy that will be filled by another H^+ ion from a neighboring site. Based on these interpretations, the H atom of the NH_4^+ from the NH_4Br is believed to interact with the oxygen in CMC (Fig. 4.4).

4.10.16.2 Carboxymethyl Carrageenan-Based Biopolymer Electrolytes

Carrageenans are a family of anionic polymers that is extracted from certain marine red algae and that share a common backbone of alternating (1→3)-linked b-D-galactopyranose and (1→4)-linked a-D-galactopyranose. There are three primary types of carrageenans: kappa carrageenan (possesses one sulfate per disaccharide), iota carrageenan (two sulfates per disaccharide), and lambda (three sulfates per disaccharide). It is extensively used in the food, pharmaceutical, and cosmetic industries as viscosity builders, gelling agents, and stabilizers. But people tried carboxymethyl carrageenan as a biopolymer electrolyte. A series of biodegradable carboxymethyl carrageenan-based polymer electrolytes, which are carboxymethyl kappa carrageenan (sulfate per disaccharide) and carboxymethyl iota carrageenan (two sulfates per disaccharide), have been studied with different ratios of lithium nitrate ($LiNO_3$) salts. The chemical modification of carrageenan derivatives (iota and kappa) for host polymer applications demonstrated that the carboxymethyl derivatives possessed conductivity values of 4.87×10^{-6} and 2.0×10^{-4} S cm^{-1}, achieved by carboxymethyl iota carrageenan and carboxymethyl kappa carrageenan, respectively. To enhance the potential use of the carboxymethyl derivatives for electrochemical applications, such as with lithium batteries, lithium nitrate ($LiNO_3$) will be introduced as a dopant salt. Due to lithium-oxygen batteries becoming increasingly attractive due to their high theoretical energy densities, proper cell voltages, and utilization of environment-friendly components, the selection of lithium nitrate is a great choice. The unique properties of the nitrate electrolyte allow it to serve as a secondary source for oxygen at the cathode as well as provide a nitrate ion pathway for oxygen reduction. The highest ionic conductivities achieved for the carboxymethyl kappa and iota carrageenan were 5.85×10^{-3} and 5.51×10^{-3} S cm^{-1} at 30 wt.% and 20 wt.% of the lithium salts addition, respectively. Cellulose-based novel biopolymer electrolytes for Na-ion batteries are reported from Francesca Colo et al. [30]

In this work, solid polymer electrolyte film is based on a polyethylene oxide (PEO) backbone, homogeneously blended with sodium carboxymethyl cellulose (Na-CMC) and sodium perchlorate. Carboxymethyl cellulose (CMC), an important ionic ether derivative of cellulose, is prepared from alkaline cellulose and chloroacetic acid by etherification and finally used as its sodium salt. Na-CMC has been selected as it can improve the mechanical integrity of the final film without affecting the ionic mobility and the electrode/electrolyte interfacial characteristics. Moreover, it shows good solubility in water, has a low cost, shows biodegradability and biocompatibility, and has a lack of toxicity. Sodium exhibits a suitable redox potential and is very abundant in the earth's crust. Such aspects of sodium-based precursors and their related economy make Na-ion batteries eco-friendly. The concerns of the electrolyte, where standard organic liquid electrolytes are mainly used, such as $NaClO_4$ or $NaPF_6$ salts dissolved in propylene carbonate (PC). Carboxymethyl cellulose (CMC), an important ionic ether derivative of cellulose that is prepared from alkaline cellulose and chloroacetic acid by etherification and usually used as its sodium salt (Na-CMC), is widely applied in industrial applications. Na-CMC has been selected as it can improve the mechanical integrity of the final film without affecting the ionic mobility and the electrode/electrolyte interfacial characteristics. It shows good solubility in water, has a low cost, shows biodegradability and biocompatibility, and has a lack of toxicity. Na-CMC has been also used as a binder for the active electrode material particles, which enables the overall process-including the electrodes and electrolyte preparation-to be carried out through very simple, cheap, and absolutely eco-friendly water-based material. The promising results are reported using sodium carboxymethyl cellulose.

4.10.17 Blend Biopolymer Electrolytes (BBPE) for Batteries

4.10.17.1 Starch-Chitosan-Based Biopolymer Electrolytes for Proton Batteries

The miscibility between starch and chitosan is confirmed by DSC analysis where they reported only one T_g at the 80 wt.% starch-20 wt.% chitosan. In most systems, proton species are contributed by the addition of salts or inorganic acids. Polymer inorganic acid complexes always suffer from chemical degradation and mechanical integrity. So, NH_4I being used as the proton provider is promising due to the low lattice energy of the salt.

4.10.17.2 Electrolytes Preparation

For the preparation of an unplasticized system, 0.80 g of corn starch was dissolved in 100 mL of 1% acetic acid while being heated at 80°C. After the solution was cooled down to room temperature, 0.20 g of chitosan was then added to the solution. The mixtures were stirred until fully dissolved before different amounts of NH_4I were added to the starch-chitosan solutions. To prepare the plasticized system, different concentrations of glycerol were added into the highest conducting electrolyte of the unplasticized system and stirred until it became a homogeneous solution. All solutions were then cast into plastic petri dishes and left to dry at room temperature. The dried films were kept in a desiccator filled with silica gel desiccants for further drying.

4.10.17.3 Transference Number for Biopolymer Electrolyte

In the biopolymer electrolyte system, different ions have different mobilities, thus carrying different portions of the total current. Both the cation and the anion have a chance to move, but in a battery only cations are responsible for the intercalation and deintercalation process at the cathode during the charge-discharge cycle. Cation transference number analysis is applicable for electrolytes with two mobile ions sandwiched between two electrodes. MnO_2 was chosen to be the reversible electrode because it is transparent to the cations and electrons, hence suitable for anions blocking. The proton transference number was determined by Watnabe's technique and calculated using:

$$t_+ = \frac{R_b}{\frac{\Delta V}{I_{ss}} - R_C} \tag{4.5}$$

where ΔV is the bias voltage from DC polarization, R_b is the bulk resistance, and R_c is the charge-transfer resistance. These values can be determined from impedance spectroscopy. The value of t_+ is found to be 0.40. Other reports show that the cation transference numbers range from 0.21 to 0.46. Although the transference number may be affected by the ion association, it still offers an insight into the ion transport process. The primary proton battery can stand up to 65 h when discharged at 0.10 mA. The internal resistance, short circuit current, and the power density of the primary proton batteries are 62.30 Ω, 17.70 mA, and 4.06 mW cm^{-2}, respectively. The secondary proton battery has been charged and discharged at 0.40 mA for 60 cycles. These results conclude that NH_4I can be a good proton provider in this biomaterial proton battery for low-current-density devices [31].

4.10.17.4 Chitosan-PEO Blend Polymer Electrolyte for Proton Batteries

Shukur et al. prepared plasticized chitosan-poly(ethylene oxide) (PEO) doped with ammonium nitrate (NH_4NO_3) electrolyte films prepared by the solution-cast technique [32].

The sample with 70 wt.% ethylene carbonate (EC) exhibits the highest room temperature conductivity of $(2.06 \pm 0.39) \times 10^{-3}$ S cm^{-1}. The proton battery is fabricated and shows an open circuit potential (OCP) of (1.66 ± 0.02) V and average discharge capacity at (48.0 ± 5.0) mA h. The maximum power density of the fabricated cell is (9.73 ± 0.75) mW cm^{-2}. The polymer electrolyte is also employed as a separator in the electrical double layer capacitor (EDLC) and is cycled for 140 times at room temperature.

4.10.17.5 Chitosan-PVA Blend Polymer Electrolyte for Proton Batteries

In this electrolyte, a 36 wt.% PVA and a 24 wt.% chitosan blend doped with 40 wt.% NH_4NO_3 exhibited the highest room temperature. The conductivity value obtained was 2.07×10^{-5} S cm^{-1}. When varying the plasticizer, the electrolyte conductivity reached 1.60×10^{-3} S cm^{-1}. The conductivity is attributed to the increase in the number of mobile ions and in the ionic mobility of the ions. The open circuit voltage of the primary battery is about 1.63 V and the secondary battery is about 1.65 V. The overall reaction in the cell described according to $Zn^+ \cdot ZnSO_4 \cdot 7H_2O$/PVA-chitosan-$NH_4NO_3$-EC/$MnO_2$ cells with no liquid electrolyte in the cathode. By adding the electrolyte solution to the cathode, the cell can be recharged and is able to perform nine cycles for almost 90 h [33].

4.10.18 Gel Biopolymer Electrolytes (GBPE) for Batteries

4.10.18.1 Xanthan and k-Carrageenan Gels

Xanthan and k-carrageenan were used to prepare hydrogels to be used as electrolytes in aluminium air primary batteries. Xanthan gum is a bacterial polysaccharide commercially produced by secretion from the bacterium *Xanthomonas campestris* in aerobic fermentation conditions. The structure consists of a cellulose backbone (β-(1→4)-D-glucose) substituted at C-3 on alternate glucose residues with a trisaccharide side-chain. It is widely used as a good food additive. The primary structure can vary with the algal source, but its backbone is essentially based on a repeating sequence of β-D-galactopyranose residues linked glycosidically through positions 1 and 3 and α-galactopyranose residues linked glycosidically through positions 1 and 4. In this work, the two above polysaccharides were used as a starting solid matrix to prepare hydrogels with alkaline solutions (KOH 1 and 8 M). These solutions are known to effectively work as liquid electrolytes in many types of electrochemical devices while the potassium counter ion was selected for its ability to enhance the gelling process of k-carrageenan. The products obtained were tested as gel polymer electrolytes in an Al/air electrochemical cell. This system was chosen for the interesting electrochemical properties of Al that theoretically could provide very high energy and power densities and compete with Li batteries for different applications, from mobile to automotive.

4.10.18.2 Preparation of Gel Biopolymer Electrolytes and Their Characterization Studies

A pasty gel was obtained by stirring xanthan in KOH solution, while a clear and effective gelling process was observed with k-carrageenan in 8 M KOH. The maximum conductivity (8.8×10^{-2} S cm^{-1}) was observed for gels prepared with the highest concentration KOH solution. Discharge tests effected on Al/air cells evidenced high values of energy delivered when gels made by 8 M solutions were used (18 and 33 mWh cm^{-2}, for xanthan and k-carrageenan-based gels, respectively.) The best performance was obtained with the gel prepared by k-carrageenan in terms of cell capacity and duration. The discharge curve shows the best cell behavior, as it reached a cell capacity of 53 mAh cm^{-2} with respect to 28 mAh cm^{-2}.

4.10.18.3 Lignin Acts as a Gel Polymer Electrolyte

The electrolyte membrane based on lignin can be easily fabricated with lignin, a liquid electrolyte, and distilled water. It was found that the liquid electrolyte uptake reaches up to 230 wt.%. GPE does not lose any weight and is thermally stable; at room temperature the ion conductivity is 3.73 mS cm^{-1}. The amazing property of the lithium ion transference number is high up to 0.85. GPE expresses complete electrochemical stability up to 7.5 V and is favorable and compatible with the lithium anode. It also showed excellent outstanding cell performance of cell rate and cycle stability. Lignin, one of the main constituents of lignocellulose biomass, is the second most abundant biopolymer on Earth. Lignin is a three-dimensional polymer consisting of 9-carbon phenol propane units, nonuniformly linked together by different types of bonds (alkyl-aryl, alkyl-alkyl and aryl-alkyl ether bonds). As a natural biopolymer, lignin is more interesting than synthetic polymers for applications in different fields due to its higher biodegradability and biocompatibility. In addition, the lignin polymer and one kind of by-product in the paper pulp manufacturing and biomass conversion process for biofuel production are readily available and cheap in nature. Every year about 50 million tons of lignin are generated all over the world, but $<10\%$ is utilized.

4.10.18.4 Preparation of Lignin-Based Electrolytes

First, 850 mg lignin was put into 40 mL distilled water in a glass cup at 35°C under constant stirring, and after 3 h, the uniform suspension was obtained. Second, the suspension was poured into a flat glass plate and the plate was heated on a heating platform to remove water at 60°C. Third, after heating for 6 h, one kind of opaque and dry membrane was produced. Fourth, the membrane was punched into round pieces with a diameter of 19 mm, and then these pieces were dried in a vacuum drying oven at 70°C for 6 h. Finally, the dried pieces were transformed into a glove box and immersed into a liquid electrolyte for enough time to obtain the GPE for further use. After preparation, the GPE was used for the cell fabrication. The cycle performance for the cell is very outstanding, and there is no evident capacity fading during cycling. The coulomb efficiencies after the 50 cycles are retained between 96% and 100%. The extreme excellent cell performance should be attributed to: (a) fairly high liquid electrolyte uptake of lignin membrane, (b) the high ionic conductivity and super high lithium ion transference number of the lignin-based electrolyte, and (c) a good electrode/electrolyte interface contact.

4.10.18.5 Carboxymethyl Cellulose

The porous membrane of carboxymethyl cellulose (CMC) from natural macromolecule is used as a host of a gel polymer electrolyte for lithium ion batteries. It is prepared by a simple nonsolvent evaporation method and its porous structure is fine adjusted. The electrolyte uptake of the porous membrane based on CMC is 75.9%. The ionic conductivity of the prepared gel membrane showed 0.48 mS cm^{-1}. The lithium ion transference in the gel membrane at room temperature is 0.46. The prepared gel membrane exhibited very good electrochemical performance, including higher reversible capacity, better rate capability, and good cycling behavior. Carboxymethyl cellulose (CMC) is a linear polymeric derivative of cellulose, consisting of β-linked glucopyranose residues with partial hydroxyl groups substituted with carboxymethyl groups. It can dissolve in water to form a viscous solution and is widely used as a binder or thickener in pharmaceuticals, foods, and ceramics.

4.10.18.6 Preparation of Carboxymethyl Cellulose

MC was dissolved in 40 mL of distilled water. To obtain suitable porosity, a variable amount of *N,N*-dimethylformamide (DMF) was added into the CMC solution under vigorous stirring at 70°C and a clear solution was prepared after further stirring for 4 h; the solution was then cast on a glass plate.

After the water was evaporated at 80°C, a thin CMC membrane with a thickness of about 20 μm was obtained. The lithium ion transference number in the gelled CMC polymer electrolytes increased with the porosity. This is due to the porous structure and the unique structure of CMC. It is known that CMC is a polyanionic cellulose that can retard the passing of PF_6 anions. As a result, more pores will provide more resistance for the movement of anions. The poly anionic structure of CMC favors the movement of lithium ions. As a result, the movement of lithium ions and the counter anion PF_6 are shown in the scheme. If the porosity can be further increased in the future, the lithium ion transference number will be higher [34].

The unusual porous structure is obtained if fine adjusted by varying the composition ratio of the solvent and nonsolvent mixture. When saturated with liquid electrolyte to form GPE, the ionic conductivity and lithium ion transference number of the GPE are higher than those of the commercial separator. It delivers excellent thermal stability because the liquid electrolyte is well retained in the CMC polymer matrix due to the existence of —OH and —COOH groups, which will markedly improve the safety of lithium ion batteries.

REFERENCES

[1] Green MA. Solar cells: operating principles, technology and system applications. Englewood Cliffs, NJ: Prentice-Hall, Inc.; 1982.

[2] Radziemska E, Klugmann E. Thermally affected parameters of the current–voltage characteristics of silicon photocell. Energy Convers Manag 2002;43:1889–900.

[3] Zhou W, Yang H, Fang Z. A novel model for photovoltaic array performance prediction. Appl Energy 2007;84:1187–98.

[4] Lenz RW. Biodegradable polymers. Adv Polym Sci 1993;107:1–40.

[5] O'Regan B, Gratzel M. A low-cost, high-efficiency solar cell based on dye-sensitized colloidal TiO_2 films. Nature 1991;353:737–40.

[6] Yum JH, Chen P, Gratzel M. Recent developments in solid-state dye-sensitized solar cells. ChemSusChem 2008;1:699.

[7] Yella A, Lee HW, Tsao HN, Yi C, Chandiran AK, Nazeeruddin MK, et al. Porphyrin-sensitized solar cells with cobalt (II/III)-based redox electrolyte exceed 12 percent efficiency. Science 2011;334:629–34.

[8] Gratzel M. Dye-sensitized solar cells. J Photochem Photobiol C 2003;4:145–53.

[9] Ripon B, I-Ming H. A SnO_2 and ZnO nanocomposite photoanodes in dye-sensitized solar cells. Electrochem Solid-State Lett 2013;2(11):Q104.

[10] Xiaoru G, Ganhua L, Junhong C. Graphene-based materials for photoanodes in dye-sensitized solar cells. Front Energy Res 2015;3.

[11] Gratzel M. Recent advances in sensitized mesoscopic solar cells. Acc Chem Res 2009;42:1788.

[12] Wu J, Lan Z, Hao S, Li P, Lin J, Huang M, et al. Progress on the electrolytes for dye-sensitized solar cells. Pure Appl Chem 2008;80:2241–58.

[13] Boschloo G, Hagfeldt A. Characteristics of the iodide/triiodide redox mediator in dye-sensitized solar cells. Acc Chem Res 2009;42:1819–26.

[14] Rudhziah S, Ahmad A, Ahmad I, Mohamed NS. Biopolymer electrolytes based on blend of kappa-carrageenan and cellulose derivatives for potential application in dye sensitized solar cell. Electrochim Acta 2015;175:162–8.

[15] Liew CW, Ramesh S, Arof AK. Good prospect of ionic liquid based-poly(vinyl alcohol) polymer electrolytes for supercapacitors with excellent electrical, electrochemical and thermal properties. Int J Hydrog Energy 2014;39:2953–63.

[16] Gong J, Sumathy K, Liang J. Polymer electrolyte based on polyethylene glycol for quasi-solid state dye sensitized solar cells. Renew Energy 2012;39:419–23.

[17] Federico B, Elena DO, Stefano B, Roberta B. Photo-polymerization of acrylic/methacrylic gel–polymer electrolyte membranes for dye-sensitized solar cells. Chem Eng J 2013;225:873–9.

[18] Wan ZNY, Wong TM, Mehboob K, Adel ES, Norani MM. Bio-based chitosan/PVdF-HFP polymer-blend for quasi-solid state electrolyte dye-sensitized solar cells. e-Polymers 2017;17:https://doi.org/10.1515/epoly-2016-0305.
[19] Mohammad HK, Ramesh S, Ramesh K. Hydroxypropyl cellulose based non-volatile gel polymer electrolytes for dye-sensitized solar cell applications using 1-methyl-3-propylimidazolium iodide ionic liquid. Sci Rep 2015;5.
[20] Finkenstadt VL. Natural polysaccharides as electroactive polymers. Appl Microbiol Biotechnol 2005;67:735–45.
[21] Selvakumar M, Krishna BD. $LiClO_4$ doped cellulose acetate as biodegradable polymer electrolyte for supercapacitors. J Appl Polym Sci 2008;110:594–602.
[22] Singh R, Baghel J, Shukla S, Bhattacharya B, Rhee HW, Singh PK. Detailed electrical measurement of sago starch biopolymer solid electrolyte. Phase Trans A Multinat J 2014;87(12):1237–45.
[23] Singh R, Baghel J, Shukla S, Bhattacharya B. Synthesis, characterization and dye sensitized solar cell fabrication using solid biopolymer electrolyte membranes. High Perform Polym 2016;28:47–54.
[24] Sudhakar YN, Selvakumar M. Lithium perchlorate doped plasticized chitosan and starch blend as biodegradable polymer electrolyte for supercapacitor. Electrochim Acta 2012;78:398–405.
[25] Selvakumar M, Krishna BD. $LiClO_4$ doped plasticized chitosan as biodegradable polymer electrolyte for supercapacitors. J Appl Polym Sci 2009;114:2445–54.
[26] Kamarulzaman N, Osman Z, Muhamad MR, Ibrahim ZA, Arof AK, Mohamed NS. Performance characteristics of $LiMn_2O_4$/polymer/carbon electrochemical cells. J Power Sources 2001;97:722–5.
[27] Sudhakar YN, Selvakumar M, Krishna BD. Effect of acid dopants in biodegradable polymer electrolyte in an electrochemical double layer capacitor. Phys Scr 2015;90.
[28] Sudhakar YN, Selvakumar M, Krishna BD. Preparation and characterization of phosphoric acid doped hydroxyethyl cellulose electrolyte for use in supercapacitor. Mater Renew Sustain Energy 2015;4:10.
[29] Gadjourova Z, Andreev YG, Tunstall DP, Bruce PG. Ionic conductivity in crystalline polymer electrolytes. Nature 2001;412:520–3.
[30] Francesca C, Federico B, RNair J, Matteo D, Claudio G. Cellulose-based novel hybrid polymer electrolytes for green and efficient Na-ion batteries. Electrochim Acta 2015;174:185–90.
[31] Yusof YM, Illias HA, Shukur MF, Kadir MFZ. Characterization of starch-chitosan blend-based electrolyte doped with ammonium iodide for application in proton batteries. Ionics 2017;23:681–97.
[32] Shukur MF, Ithnin R, Illias HA, Kadir MFZ. Proton conducting polymer electrolyte based on plasticized chitosan–PEO blend and application in electrochemical devices. Opt Mater 2013;35:1834–41.
[33] Kadir MFZ, Majid SR, Arof AK. Plasticized chitosan–PVA blend polymer electrolyte based proton battery. Electrochim Acta 2010;55:1475–82.
[34] Zhu YS, Xiao SY, Li MX, Chang Z, Wang FX, Gao J, et al. Natural macromolecule based carboxymethyl cellulose as a gel polymer electrolyte with adjustable porosity for lithium ion batteries. J Power Sources 2015;288:368–75.

Chapter 5

Biopolymer Electrolytes for Fuel Cell Applications

Chapter Outline

5.1 INTRODUCTION

Fuel cells are environmentally friendly electrochemical devices used for the conversion of chemical energy obtained from a redox reaction directly into electrical energy. Fuel cells as zero-emission power generators with high efficiency, high energy density, and quiet operations are one of the most promising candidates in this era [1,2]. They consist of an electrolyte material packed between two thin porous electrodes. The electrolyte that allows the flow of ions between the anode and cathode should be highly resistive to the electron current. The fuel and oxidant (e.g., oxygen) supplied externally passes over the anode and cathode, respectively, where they are dissociated catalytically into ions and electrons. The overall chemical reaction is the same as the burning of fuel except that in this case the reactants are separated spatially. The fuel cell intercepts the stream of electrons that spontaneously flows from the fuel to the oxidant and makes it pass through an external electrical circuit to generate power. On the other hand, the ions move toward the oppositely charged electrode through the electrolyte, the ion migration process driven fundamentally

Biopolymer Electrolytes. https://doi.org/10.1016/B978-0-12-813447-4.00005-4

by the concentration gradient between the two interfaces (electrode-electrolyte) [1,3]. Unlike in the conventional batter, the fuel and oxidant are not integral parts of a fuel cell but they are supplied as required to provide power to the external load. As long as there is a supply of the fuel to the cell, it is considered "charged" and hence, self-discharge is absent.

Development of fuel cells as a power source was attempted many years back, mainly for space and defense applications. In the 1980s and 1990s, Earth-based systems were developed [4]. Recently, the quest for more efficient and environmentally friendly electrical generation technologies has led to substantial resources being diverted to the development of fuel cells. Though commercialization of fuel cell technology is moving forward, a good amount of work still needs to be done in many fields. Development and optimization of component materials have to be done to improve the performance and lower the cost for commercial applications. Long-term testing to obtain statistical data on fuel cell performance has to be done. Development of new in situ measurement techniques for characterization of fuel cells in large-scale applications has to be looked into as fuel cell research moves on from the laboratory-scale, single-cell studies to the development of application-ready stacks [3].

5.2 OPERATING PRINCIPLE OF FUEL CELLS

A fuel cell consists of a thin composite structure of two electrodes with an electrolyte separating them (Fig. 5.1). In a fuel cell, the electrocatalysts are

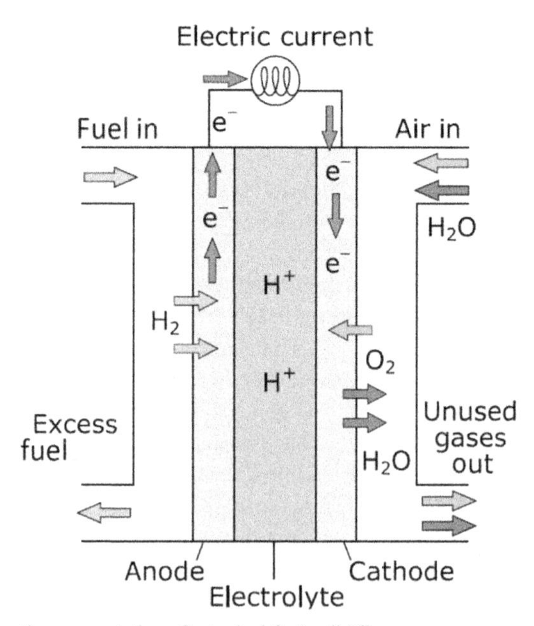

FIG. 5.1 Schematic representation of a typical fuel cell [6].

positioned on either side of an electrolyte, sometimes embedded in the porous electrodes. They can be either a polymer or ceramic material or an immobilized acid or alkali, forming the cell assembly. Effective electrocatalysts are required for good electrochemical performance of the fuel cell. To ensure a uniform performance of both anode and cathode, the distribution of reactant gases ideally has to be uniform across its surface. The fuel and oxidant are fed to the back faces of the electrodes with the flow fields controlling the supply and distribution [5]. The flow field allows the fuel and oxidant to flow along the length of the anode and cathode while permitting gas transport to the electrocatalyst through a diffusion layer, normal to its surface. The chemical energy contained within the fuel is converted by the fuel cell to electrical energy by virtue of the electrochemical reactions in the cell. The best performance in terms of power output is achieved with the reaction of pure hydrogen with oxygen, but a range of other fuels such as methanol and ethanol are also being used as fuels.

When an acid electrolyte is used in the fuel cell, the reactions taking place are as given below in Eqs. (5.1), (5.2):

$$H_2 \rightarrow 2H^+ + 2e^- \quad E^\circ = 0\,V \tag{5.1}$$

$$0.5O_2 + 2H^+ + 2e^- \rightarrow H_2O \quad E^\circ = 1.229\,V \tag{5.2}$$

Thus for this fuel cell, the theoretical standard cell potential is ~1.23 V. In the case of the combustion engine, the heat of combustion of the fuel is first converted into mechanical energy and then later converted into electrical energy. The major advantage of a fuel cell is the direct conversion of the chemical energy into electrical energy, which results in high efficiency as it is not limited by the Carnot rule. The maximum (theoretical) electrical efficiency of a fuel cell, ε, at standard conditions of temperature and pressure, is given by Eq. (5.3):

$$\varepsilon = \frac{\Delta G}{\Delta H} = \frac{-2FE^\circ}{\Delta H} \tag{5.3}$$

where $\triangle G$ is the Gibbs free energy change, $\triangle H$ is the enthalpy change of the reaction, F is Faraday's constant, and E° is the standard cell potential. A fuel cell in which the product is liquid water at low temperature has an efficiency of ~83%. In fact, due to the irreversible voltage losses associated with the flow of current and the nature of the cell constructed in almost all practical fuel cells, the theoretical efficiency is not achieved. For a fuel cell, the electrochemical efficiency, ε_e, is given by Eq. (5.4):

$$\varepsilon_e = \frac{-2FE}{\Delta G} \tag{5.4}$$

where E is the actual potential produced by the fuel cell. The voltage losses caused due to electrode polarization, internal and external cell resistances, mass transport limitations, and limitations in cell materials typically decrease the potential produced by the fuel cell with the current drawn from the cell [5]. Electrode polarization arises when current flows and causes overpotential. Overpotential is the deviation of the electrode potential from the standard potential, corresponding to the electrical work performed by the cell. In the hydrogen fuel cell, the reduction of oxygen (Eq. 5.2) is kinetically much slower than the oxidation of hydrogen. This results in the largest voltage loss. Ohmic voltage losses in the fuel cell are caused due to the two kinds of internal resistances. The first type is due to the resistance of the electrolyte that is used to ionically connect the electrode reactions. The second one is associated with the connection of the electrode material in the cell, namely, the electronic resistance. Factors like low partial pressures of oxygen in the presence of nitrogen in the air, water vapor formed due to the cathodic reaction, covering of the electrocatalysts by the liquid water, and flooding of the porous structure of the electrode makes the reactants not be able to diffuse to the electrocatalysts at the required rate of the electrochemical reaction. These mass transport limitations cause a rapid fall in potential at high currents. With an increase in the current density, the power density produced by a fuel cell exhibits a maximum. This peak power is usually used to characterize and compare the performance of the fuel cells.

5.3 IMPORTANCE OF FUEL CELLS

The use of fossil fuels has led to harmful environmental impacts due to the increasing level of pollutants and the geopolitical consequences. Fuel cells with their pollution-free operation have emerged as potential alternatives to combustion engines and are considered as an environment-benign technology due to their high efficiency, high energy density, low noise, and low maintenance costs [7–9]. Reserves of fossil fuels are limited, and it has been predicted that the production of fossil fuels will decline after peaking around 2020 [10]. The reported estimate of worldwide environmental damage due to the burning of fossil fuels is around $5 trillion annually. To satisfy the growing demand for food and energy due to population explosion, research and development in the area of fuel cells needs to be enhanced [11,12]. Fuel cells are electrochemical devices in which the electrochemical processes are not governed by Carnot's rule and hence, their operation is simple and more efficient compared to internal combustion engines. Their high efficiency makes them an attractive option for a wide range of applications, including transportation as well as stationary and portable electronic devices such as consumer electronics, laptops, video cameras, etc.

5.4 MEMBRANES USED IN FUEL CELLS

In the 1950s, ion-exchange membranes were invented as an electrolyte for fuel cells. Since then, researchers have been challenged to find the ideal membrane material as well as to optimize membrane properties for application in fuel cells to withstand the harsh operating environment. The desired properties for a membrane to be used in a fuel cell, which play an important role in the overall performance of the fuel cell, include the following:

(a) Compatible chemical properties: It has to satisfy the bonding requirements of the membrane electrode assembly and should facilitate rapid electrode kinetics.
(b) Stability: It should be mechanically and electrochemically stable in the operating environment, namely, a high resistance to oxidation, reduction, and hydrolysis. It should have high thermal stability and high durability.
(c) Transport properties: Because hydration of the membrane plays a crucial role, to maintain uniform water content it should have high water transport while also preventing localized drying. It should have good water uptakes at high temperatures of about 100°C. It should be resistant and have extremely low permeability to reactant gases to minimize columbic inefficiency. In a direct-methanol fuel cell (DMFC), this is a concern as methanol crossover takes place and gets oxidized at the cathode, reducing the cell voltage by the formation of mixed potential at the cathode.
(d) Conductivity: It should have high proton conductivity with minimal resistive losses to support high currents and have zero electronic conductivity.
(e) Type of fuel: It should be flexible to operate with a wide variety of fuels.
(f) Economic: The production costs should be compatible with the commercial requirements of the fuel cell.

5.5 CLASSIFICATION OF FUEL CELLS

Based on the type of electrolyte they use, fuel cells are classified into the following main types: alkaline fuel cell (AFC), molten carbonate fuel cell (MCFC), phosphoric acid fuel cell (PAFC), proton exchange membrane fuel cell (PEMFC), solid oxide fuel cell (SOFC), and biofuel cell [13,14]. These fuel cells use hydrogen as fuel. The DMFC, the enzymatic fuel cell (EFC), and the microbial fuel cell (MFC) are non-H_2 fuel cells in which other chemical compounds such as methanol and ethanol act as the electron donor in the anode [15,16]. A brief account of these fuel cell types has been given below.

5.5.1 Alkaline Fuel Cell (AFC)

AFCs are hydrogen-based fuel cells that are used for both the production of electricity and potable water. As the name suggests, AFCs use a liquid electrolyte solution of potassium hydroxide (KOH) due to its high alkaline hydroxide conductibility [13]. Hydroxyl ion (OH^-) from potassium hydroxide migrates toward the anode from the cathode (Fig. 5.2). At the anode, electrons are

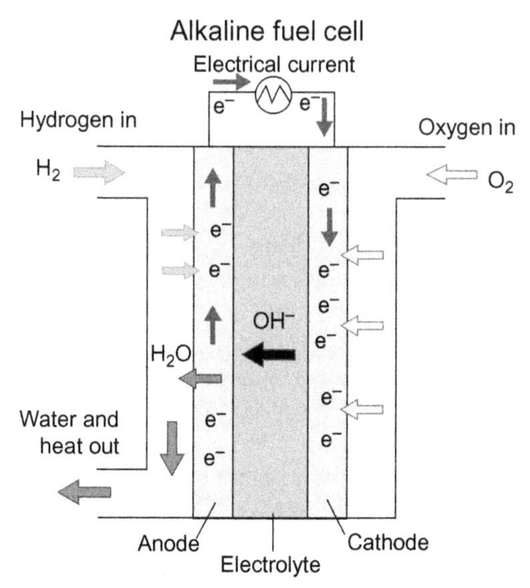

FIG. 5.2 Schematic representation of an AFC [1].

released and water is produced due to the reaction of hydrogen gas with the OH^- ions [15,16]. The overall reactions are given in Eqs. (5.5)–(5.7):

$$\text{Anode}: 2H_2 + 4OH^- \rightarrow 4H_2O + 4e^- \quad (5.5)$$

$$\text{Cathode}: O_2 + 2H_2O + 4e^- \rightarrow 4OH^- \quad (5.6)$$

$$\text{Overall cell reaction}: 2H_2 + O_2 \rightarrow 2H_2O + \text{electrical energy} + \text{heat} \quad (5.7)$$

AFCs have an advantage over other fuel cells in the fact that they can operate at flexible temperatures. While high-temperature AFCs operate at temperatures between 100°C and 250°C, the newer designs operate at much lower temperatures of 23–70°C. Also, the higher reaction kinetics at the electrodes result in higher cell voltages in AFCs. Due to high electrical efficiency, the use of lower quantities of a noble expensive metal catalyst such as platinum is feasible [13]. They have been designed commonly for transport applications [17]. They are the first and the only type of fuel cells routinely and successfully applied in space exploration, including space shuttle missions in the United States [1]. The liquid electrolyte KOH solution used is very sensitive to the presence of CO_2 and hence AFCs need very pure reactants, which is a major drawback [13]. If air is used instead of oxygen, the KOH reacts with the CO_2 in the air to form solid K_2CO_3, interfering with other reactions within the fuel cell [18]. AFCs also need a highly specific amount of the liquid electrolyte, as anything higher or lower can cause electrode flooding or electrode drying [13].

5.5.2 Molten Carbonate Fuel Cell (MCFC)

MCFCs are fuel cells that operate at high temperatures using an electrolyte composed of a molten carbonate salt mixture (salt of sodium or magnesium carbonate) suspended in a porous, chemically inert ceramic lithium aluminum oxide ($LiAlO_2$) matrix (Fig. 5.3). When heated to 650°C, the electrolytes in MCFCs melt and conduct carbonate ions (CO_3^{2-}) from the cathode to the anode [19]. At the anode, hydrogen combines with carbonate ions, producing water and carbon dioxide and releasing electrons to the external circuit. At the cathode, oxygen is reduced by carbon dioxide and electrons to carbonate ions [1]. The reactions taking place within the fuel cell are given in Eqs. (5.8)–(5.10):

$$\text{Anode}: H_2 + CO_3^{2-} \rightarrow CO_2 + H_2O + 2e^- \tag{5.8}$$

$$\text{Cathode}: \frac{1}{2}O_2 + CO_2 + 2e^- \rightarrow CO_3^{2-} \tag{5.9}$$

$$\text{Overall cell reaction}: H_2 + \frac{1}{2}O_2 \rightarrow H_2O + \text{electrical energy} + \text{heat} \tag{5.10}$$

The advantage of MCFCs comes from the fact that the high operating temperature limits damage from the carbon monoxide poisoning of the cells and also waste heat generated can be recycled to make additional electricity [20]. High operating temperatures and the use of corrosive electrolytes in MCFCs

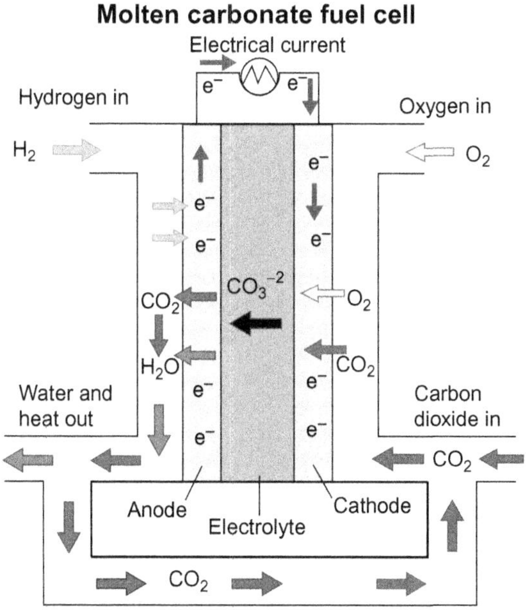

FIG. 5.3 Schematic representation of an MCFC [1].

accelerate component breakdown and cell life, hence reduced durability becomes a major problem. Development of corrosion-resistant materials for fuel cell components to increase cell life without decreasing the performance can help in overcoming the disadvantage [1].

5.5.3 Phosphoric Acid Fuel Cell (PAFC)

PAFCs are the first generation of commercial modern fuel cells operating at temperatures between 170°C and 210°C that were developed for terrestrial applications, unlike AFCs that were targeted for spacecraft. As the name suggests, the PAFCs use phosphoric acid (H_3PO_4) in highly concentrated form (>95%) as its electrolyte and porous carbon electrodes contain the platinum catalyst (Fig. 5.4). CO_2 containing air is used as the oxidant and pure hydrogen or a hydrogen-rich gas acts as the primary fuel for power generation [1]. A porous silicon-carbide matrix is often used to immobilize the electrolyte by capillary action. Due to the acid electrolyte used, the internal parts of PAFCs must be able to withstand high corrosive environments. Also, they can tolerate about 1.5% carbon monoxide concentration and hence are widely used [21]. Because the electrolyte used is tolerant to the presence of CO_2 in the fuel streams, hydrogen produced by steam reforming of organic fuels, such as hydrocarbons (typically natural gas or methane) and alcohols (mainly methanol or ethanol), are

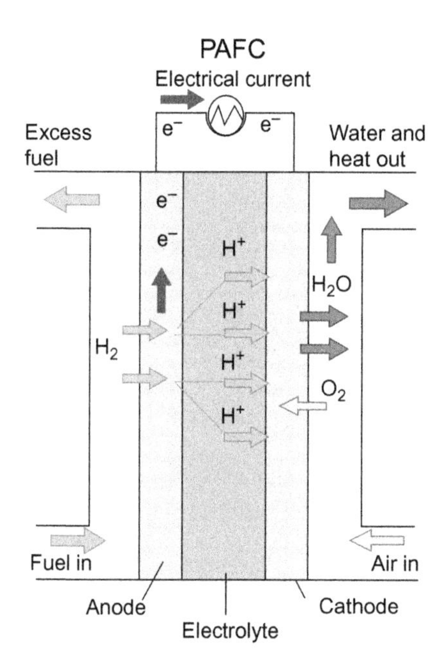

FIG. 5.4 Schematic representation of a PAFC [1].

often used as the anodic reactants [1]. PAFCs when used for the cogeneration of electricity and heat are more efficient (80%) as compared to generating electricity alone (40%–50%) [22].

5.5.4 Proton Exchange Membrane Fuel Cell (PEMFC)

PEMFCs, also known as polymer electrolyte membrane fuel cells, consist of a porous carbon electrode containing platinum or ruthenium catalyst and a proton conducting cast in solid polymer form (Fig. 5.5). The solid polymer electrolyte membrane, which is the most important component of a PEMFC, is typically based on Nafion, sulfonated polyether-ether ketone or similar polymers that allow protons to pass through but not electrons. PEMFCs operating at low temperatures, $<80°C$, have the advantage of being lightweight, delivering high-power density, and being free from corrosive liquid; this makes it have a longer cell lifetime compared to other fuel cells [1]. PEMFCs have high energy efficiency, which makes them useful in a variety of applications such as transportation as well as stationary and portable power. There are two types of PEMFCs: hydrogen fuel cell (HFC) and the DMFC, both of which utilize a proton exchange membrane for the transfer of protons [17]. When hydrogen is being

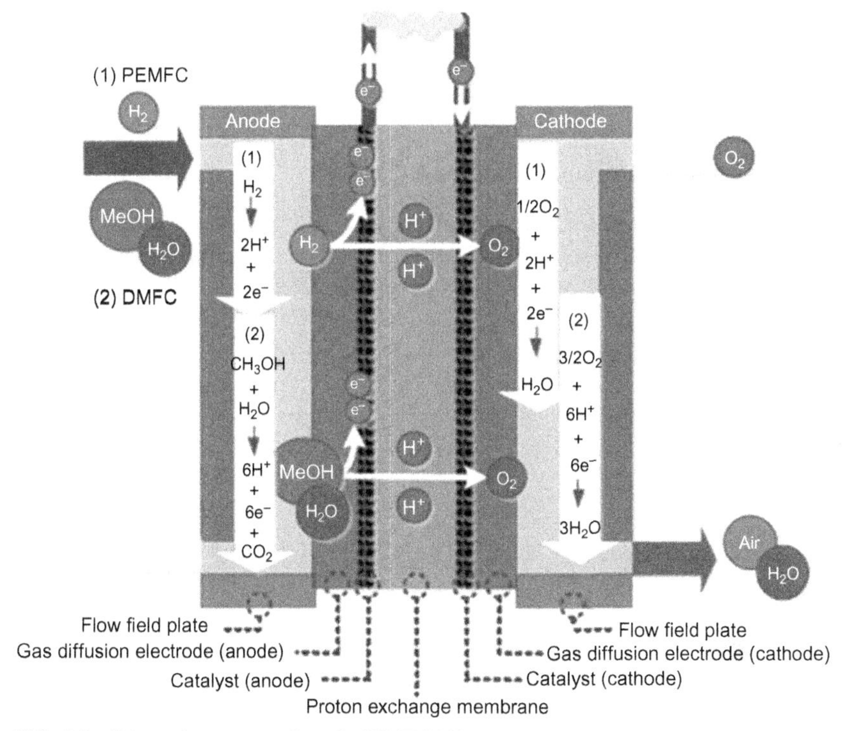

FIG. 5.5 Schematic representation of a PEMFC [3].

used as a fuel, a separate reformer reactor is required that is not required in the case where methanol is used as a fuel. In HFCs, hydrogen is oxidized at the anode to liberate two electrons that travel through the electronic circuits and two protons that travel through the proton exchange membrane [23]. At the cathode, water is produced due to the reduction of oxygen. The overall reactions taking place within the fuel cell are given in Eqs. (5.11)–(5.13):

$$\text{Anode}: H_2 \rightarrow 2H^+ + 2e^- \tag{5.11}$$

$$\text{Cathode}: \frac{1}{2}O_2 + 2H^+ + 2e^- \rightarrow H_2O \tag{5.12}$$

$$\text{Overall cell reaction}: H_2 + \frac{1}{2}O_2 \rightarrow H_2O \tag{5.13}$$

The polymer membrane in a PEMFC carries out two major functions. It acts as an electrolyte providing ionic communication between the electrodes and it acts a separator for the reactants. They amount to up to 30% of the material cost of the entire fuel cell and exhibit shortcomings that result in poor performance of the cell. Optimized proton and water transport properties of the membrane are essential as dehydration will reduce the proton conductivity and excess water can flood the electrodes. To improve the water retention properties, various approaches have been made such as modifying perfluorinated ionomer membranes, the functionalization of aromatic hydrocarbon polymers/membranes, the development of composite membranes based on solid inorganic proton-conducting materials, and an organic polymer matrix.

5.5.5 Direct Methanol Fuel Cell (DMFC)

DMFC is a type of PEMFC that uses methanol as the fuel instead of hydrogen or hydrogen-rich gas. The solution of methanol and water is internally reformed by the catalyst and oxidized at the anode to liberate electrons and protons (Fig. 5.6). The cathode reaction for a DMFC is similar to a HFC [24]. The overall reactions taking place in DMFC are given in Eqs. (5.14)–(5.16):

$$\text{Anode}: CH_3OH\,(l) + H_2O\,(l) \xrightarrow{Pt/Ru} CO_2(g) + 6H^+ + 6e^- \tag{5.14}$$

$$\text{Cathode}: \frac{3}{2}O_2(g) + 6H^+ + 6e^- \xrightarrow{pt} 3H_2O(l) \tag{5.15}$$

$$\text{Overall cell reaction}: CH_3OH\,(l) + \frac{3}{2}O_2(g) \rightarrow CO_2(g) + 2H_2O(l) \tag{5.16}$$

The DMFC has several advantages as a promising energy source over HFC due to its ease of fuel delivery, storage, low cost of methanol, operation at low temperature and pressure, lack of humidification requirements, reduced design complexity, and high power density [26]. Because fuel cell efficiency and power density strongly depend on the conductance of electrolytes, only acidic

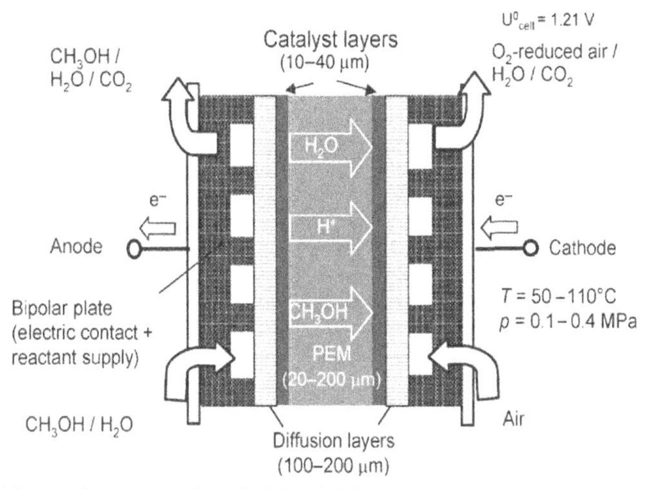

FIG. 5.6 Schematic representation of a DMFC [25].

electrolytes can be used in a DMFC to aid the rejection of carbon dioxide. Also limited operating temperatures, susceptibility to osmotic swelling, methanol crossover, and the high costs of equipment act as factors that hinder the commercialization of DMFC [27].

5.5.6 Solid Oxide Fuel Cell (SOFC)

SOFCs are fuel cells operating at very high temperatures of about 1000°C applied in large-scale stationary power generators to provide electricity for factories and towns [15]. A hard ceramic compound of metal, such as calcium oxide or zirconium oxide, is used as the electrolyte with hydrogen and carbon monoxide acting as the reactive fuels (Fig. 5.7). A cathode oxygen is supplied in the form of air [28]. Due to the high operating temperature of SOFCs, the negatively charged oxygen ions migrate through the crystal lattice. When the fuel passes over the anode, it gets oxidized by the oxygen ions, which react with them after moving across the electrolyte. SOFCs are generally 50%–60% efficient, but with cogeneration by utilization of waste heat the efficiency could increase to 80%–85%. The different reactions taking place within the fuel cell are given in Eqs. (5.17)–(5.19) when hydrogen is used as fuel and Eqs. (5.17), (5.20), (5.21) when carbon monoxide is used as fuel [28]

$$\text{Cathode}: \frac{1}{2}O_2 + 2e^- \rightarrow O^{2-} \tag{5.17}$$

$$\text{Anode}: H_2 + O^{2-} \rightarrow H_2O + 2e^- \tag{5.18}$$

$$\text{Overall cell reaction}: H_2 + \frac{1}{2}O_2 \rightarrow H_2O + \text{electrical energy} + \text{heat} \tag{5.19}$$

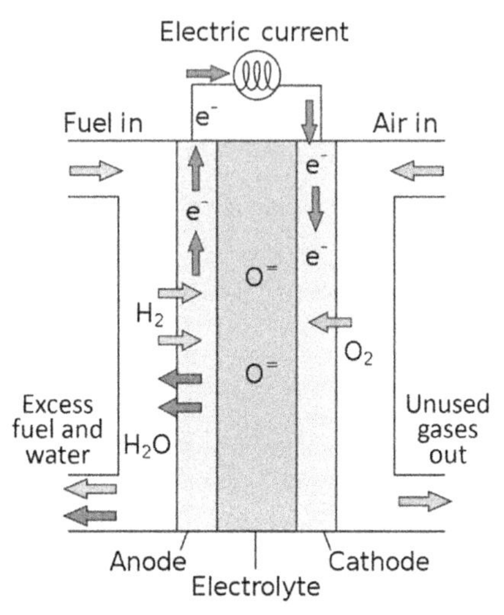

FIG. 5.7 Schematic representation of an SOFC [1].

$$\text{Anode}: CO + O^{2-} \rightarrow CO_2 + 2e^- \tag{5.20}$$

$$\text{Overall cell reaction}: \ CO + \frac{1}{2}O_2 \rightarrow CO_2 \tag{5.21}$$

SOFC has the advantage of being low cost due to the high temperature operation, which eliminates the need for a precious-metal catalyst. But it suffers from the drawback that high temperature limits applications of SOFCs, the units tend to be large in size, and the solid electrolytes can undergo cracking, although leakage is not an issue [29].

5.5.7 Biofuel Cell

Microorganisms have the ability to extract energy from a range of chemical substrates, which led to them being used in power generation systems. Biofuel cells are a category of fuel cells that employs biocatalysts such as a microorganism, enzyme, or even an organelle interacting with an electrode surface [4]. Like other fuel cells, they contain porous electrodes to support fuel transport to the catalyst reaction sites and a polymer electrolyte membrane or a salt bridge to separate the electrodes. Based on the type of biocatalyst involved, they are classified into enzymatic fuel cells (EFC) and microbial fuel cells (MFC). Biofuel cells offer substantial benefits in terms of catalytic activity, specificity, and cost [30].

EFCs in the early stage of their development have replaced expensive metal catalysts with cheap enzymes, which have the advantage of specificity and hence eliminate the use of a membrane separator. EFCS have several drawbacks such as a limited lifetime (typically 7–10 days) due to the fragile nature of enzymes and a lower efficiency due to the partial oxidation of fuel due to the employment of a single type of enzyme. To increase the power density of EFCs more than the conventional fuel cell, anodes should possess multidimensional and multidirectional pore structures to optimize the need for surface area. It must support an efficient charge transfer mechanism and balance the electron transfer with the proton transfer. The multienzyme system should be successfully immobilized to completely oxidize the fuel to carbon dioxide. The two main application areas that are being considered for enzymatic biofuel cell EFCS are currently being applied in in vivo implantable power supplies for sensors and pacemakers and ex vivo power supplies for small portable power devices (wireless sensor networks, portable electronics, etc.) [30]. Several enzymes such as glucose oxidase and dehydrogenase, dehydrogenase enzymes for alcohol, formate, lactic acid, and formaldehyde are successfully immobilized and used in EFCs to date.

MFCs employ living cells such as electrogenic bacteria in the place of a catalyst layer to catalyze the oxidation of fuels. In MFCs, the fuels such as simple sugars are completely oxidized to carbon dioxide by microbes and in the process transfer the electrons to the anode. At the cathode, oxygen is abiotically or biotically reduced by the electrons to produce water. The positive charges migrate from anode to cathode through an ion-permeable separator to maintain the charge neutrality. MFCs offer a promising technology for efficient wastewater treatment. MFCs generate energy as direct electricity for onsite remote application and are long-lived systems (life span of up to 5 years). But they suffer from the disadvantage of low power densities due to slow transport across cellular membranes.

5.6 CHALLENGES IN FUEL CELL TECHNOLOGY

Despite the research and improvements in fuel cell design and components made over the past several years, many issues have to be addressed before fuel cells can become competitive enough to be used commercially. The most important hurdle is the fuel. Fuel cells like PEMFCs run best on pure hydrogen, but the onboard storage of hydrogen is complicated. Also, hydrogen obtained from hydrocarbons contains small amounts of carbon monoxide, which have disastrous effects on the efficiency of the anode reaction. The efficiency of fuel cells like DMFCs with methanol as the fuel is also reduced due to the complicated system of onboard reformation of methanol. It suffers from poor kinetics of methanol oxidation reaction at low temperatures and a high methanol permeation rate through the membrane. The overall cost of the fuel cells with expensive metal-based catalysts and problems with the stability of the system thwart their commercialization.

Interest in certain types of fuel cell technologies decreased due to economic factors, material problems, and certain inadequacies in the operation of these electrochemical devices. Because the use of liquid electrolytes in the fuel cell acts as a drawback, polymer electrolytes can act as an alternative, which also eliminates the need to use pure fuels [18,19]. Membranes used in fuel cells have to meet several desired properties such as high proton conductivity, low electronic conductivity, impermeability to fuel, good mechanical strength in both the dry and hydrated states, and high oxidative and hydrolytic stability in a real environment, which depends on properties such as ion exchange capacity, morphology, and water uptake. This must be assessed when characterizing the potential of a new fuel cell membrane [11,13].

5.7 BIOPOLYMER ELECTROLYTES FOR FUEL CELL APPLICATIONS

Chitosan is a derivative of chitin, a naturally occurring biopolymer. It is receiving great interest as material for both the membrane electrolyte and the electrode in various fuel cells such as polymer electrolyte-based fuel cells, including alkaline polymer electrolyte fuel cells, direct methanol fuel cells, and biofuel cells [12,31]. Considering that the polymer membrane electrolyte is the core and most expensive component of a polymer electrolyte-based fuel cell, the use of a low-cost chitosan-based membrane with desirable properties might bring down the cost [12]. Chitosan has excellent biocompatibility, nontoxicity, and chemical and thermal stability. Chitosan modified by processes such as sulfonation, phosphorylation, quaternization, and chemical cross-linking will make a cost-effective polymer electrolyte membrane with low methanol permeability and suitable ion conductivity by generating ion exchange sites on the backbone, especially at high temperature. But care should be taken that it does not affect the mechanical strength as these processes tend to increase its solubility in an aqueous medium [10]. Chitosan has an advantage over other polysaccharides due to the presence of glucosamine residues. It can form polyelectrolyte complexes due to the positive charge arising out of highly protonated amino functionalities, with a wide variety of negatively charged polyanions such as lipids, collagen, glycosaminoglycans, lignosulfonate, and alginate as well as charged synthetic polymers and DNA through electrostatic interaction [25].

An easily modifiable blend membrane consisting of chitosan and sodium alginate biopolymers forming a poly ion complex with low methanol permeability, high mechanical strength, and high proton conductivity has been used for fuel cell applications due to their abundance in nature and low cost [2,32]. They are particularly useful in the low to intermediate temperature range. To overcome the low mechanical strength of alginate due to its hydrophilic nature, inorganic fillers or graphene oxide (GO) nanocomposites can be used. Hydrogen bonding and high interfacial adhesion between the GO filler and the alginate matrix enhance the thermal and mechanical stability [2]. Polyelectrolyte membranes of alginate and carrageenan mixtures that are crosslinked and sulfonated show lesser methanol permeability and higher proton conductivity with an increase in carrageenan concentration. The aim is to increase the proton conductivity of the biopolymer electrolyte to the levels of Nafion membranes.

A stabilized silicotungstic acid-chitosan-polyvinyl alcohol membrane has higher conductivity than the Nafion. Chitosan-based membranes, both anionic and cationic, in combination with Nafion is used to enhance methanol resistance of Nafion and proton conductivity of chitosan.

5.8 CONCLUSION

The fuel cell is regarded as one of the promising technologies for future advanced energy generation. Since the discovery of the first fuel cell, progress in developing fuel cell technology for commercialization has been steady if not spectacular and is still the subject on ongoing research. The most significant progress has been in the field of the PEMFC system for transport applications, providing the basis for a low-temperature stationary unit for combined heat and power applications. Special attention has been given to composite techniques in developing membranes for the fuel cell. Among the electrolyte materials, solid polymer-based electrolyte membranes are a go-to material. However, due to the expensive nature of these materials, cost effective eco-friendly biopolymer electrolytes from renewable sources are sought after. Due to the abundance of polysaccharides in nature, they are the best candidates among natural polymers. Recently, a lot of effort has been made in the utilization of such biopolymer electrolytes and their composites with improved properties for application in fuel cells. The future is very bright for biopolymer electrolytes.

REFERENCES

[1] Xianguo L. Principles of fuel cells. New York, London: Taylor and Francis; 1962.

[2] Ma J, Sahai Y. Chitosan biopolymer for fuel cell applications. Carbohydr Polym 2013;92:955–75.

[3] Ye YS, Rick J, Hwang BJ. Water soluble polymers as proton exchange membranes for fuel cells. Polymer 2012;4:913–63.

[4] Vaghari H, Jafarizadeh-Malmiri H, Berenjian A, Anarjan N. Recent advances in application of chitosan in fuel cells. Sustain Chem Process 2013;1:16.

[5] Scott K, Shukla AK. Polymer electrolyte membrane fuel cells: principles and advances. Rev Environ Sci Biotechnol 2004;3:273–80.

[6] www.ballard.org.

[7] SMJ Z, Matsuraa T, editors. Polymer membranes for fuel cells. New York: Springer; 2009.

[8] Yan Q, Toghiani H, Causey H. Steady state and dynamic performance of proton exchange membrane fuel cells (PEMFCs) under various operating conditions and load changes. J Power Sources 2006;161:492–502.

[9] Smitha B, Sridhar S, Khan AA. Solid polymer electrolyte membranes for fuel cell applications-a review. J Membr Sci 2005;259:10–26.

[10] Sopian, K, Shamsuddin AH, Veziroglu TN, Solar hydrogen energy option for Malaysia Proceeding of the international conference on advances in stratic technology, June 1995, *UKM, Bangi* 1995:209 – 220.

[11] Kordesch KV, Simader GR. Environmental impact of fuel cell technology. Chem Rev 1995;95:91–207.

[12] Barbir F. PEM fuel cells, theory and practice. California: Elsevier Academic Press; 2005.
[13] Merle G, Wessling M, Nijmeijer K. Anion exchange membranes for alkaline fuel cells: a review. J Membr Sci 2011;377:1–35.
[14] Winter M, Brodd RJ. What are batteries, fuel cells, and supercapacitors? Chem Rev 2004;104:4245–70.
[15] Cooper HW. A future in fuel cells. Chem Eng Prog 2007;103:34–43.
[16] Sopian K, Wan Daud RW. Challenges and future developments in proton exchange membrane fuel cells. Renew Energy 2006;35:719–27.
[17] Odeh AO, Osifo P, Noemagus H. Chitosan: a low cost material for the production of membrane for use in PEMFC-A review. Energ Sources Part A 2013;35:152–63.
[18] Gülzow E, Schulze M. Long-term operation of AFC electrodes with CO_2 containing gases. J Power Sources 2004;127:243–51.
[19] Bischoff M. A high temperature fuel cell on the edge to commercialization. J Power Sources 2006;160:842–5.
[20] Amorelli A, Wilkinson MB, Bedont P, Capobianco P, Marcenaro B, Parodi F, et al. An experimental investigation into the use of molten carbonate fuel cells to capture CO2 from gas turbine exhaust gases. Energy 2004;29:1279–84.
[21] Sammes N, Bove R, Stahl K. Phosphoric acid fuel cell: fundamentals and applications. Curr Opin Solid State Mater Sci 2004;8:372–8.
[22] Acres GJK. Recent advances in fuel cells technology and its applications. J Power Sources 2001;100:60–6.
[23] Hoogers G, editor. Fuel cell technology handbook. Boca Raton, FL: CRC Press; 2003. p. 8–39.
[24] Kreuer KD. On the development of proton conducting polymer membranes for hydrogen and methanol fuel cells. J Membr Sci 2001;185:29–39.
[25] Sundmacher K, Rihko-Struckmann LK, Galvita V. Solid electrolyte membrane reactors: Status and trends. Catal Today 2005;104(2–4):185–99.
[26] Othman MHD, Ismail AF, Mustafa A. Recent development of polymer electrolyte membranes for direct methanol fuel cell application—a review. Malaysian Polym J 2010;5:1–36.
[27] Bagotsky VS. In: Bagotsky VS, editor. Fuel cells: problems and solutions. Hobken, NJ: John Wiley and Sons Inc.; 2009. p. 45–70.
[28] Offer GJ, Brandon NP. The effect of current density and temperature on the degradation of nickel cermet electrodes by carbon monoxide in solid oxide fuel cells. Chem Eng Sci 2009;64:2291–300.
[29] Scott K, Yu EH, Ghangrekar MM, Erable B, Duteanu NM. Biological and microbial fuel cells. Compr Renew Energy 2012;4:277–300.
[30] Shaari N, Kamarudin SK. Chitosan and alginate types of bio-membrane in fuel cell application: an overview. J Power Sources 2015;289:71–80.
[31] Varshney P, Gupta S. Natural polymer-based electrolytes for electrochemical devices: a review. Ionics 2011;17:479–83.
[32] Pasini Cabello SD, Moll S, Ochoa NA, Marchese J, Gim E, Compan V. New bio-polymeric membranes composed of alginate-carrageenan to be applied as polymer electrolyte membranes for DMFC. J Power Sources 2014;345–55.

Chapter 6

Biopolymer Degradation

Chapter Outline

6.1 INTRODUCTION

Biopolymers can be classified into biodegradable and nonbiodegradable biopolymers such as aliphatic polyesters and polyphosphoester (PPE). Hence, a distinction has to be made between degradability (mechanical disintegration) and biodegradability (metabolism). Degradation means the polymers disintegrate into smaller pieces biologically due to wear and tear (mechanical degradation), sunlight (photodegradation), heat (thermal degradation), ionizing radiation (radiodegradation), etc. Polymers subjected to oxygen are degraded much faster in the presence of radiation than in its absence and vice versa. Most synthetic biopolymers undergo chemical modification upon irradiation with UV light because they or their additives have chromophoric groups [1].

The doped salts/acids/bases/metal oxides in biopolymer electrolytes mainly act as prodegradants, which enhance primary photolytic and thermal degradation, bringing down the molecular weight of the biopolymer to <1000. Hence, prodegradants promote secondary biodegradation by microbes and enhance biodegradation rates.

Biodegradability is a certified performance characteristic (ISO 17088, EN 13432 in EU, ASTM D 6400 in North America, GreenPla in Japan); a few are shown in Table 6.1 [9]. To fulfill, for example, the EN 13432 norm, the polymer has to be converted to CO_2 by more than 90% within 180 days under defined conditions of humidity, temperature, and oxygen. Biopolymers from natural origins such as lignin, starch, cellulose, and chitin as well as synthetic biopolymers such as poly(caprolactone), poly(vinyl alcohol), poly(lactic acid),

Biopolymer Electrolytes. https://doi.org/10.1016/B978-0-12-813447-4.00006-6

TABLE 6.1 A Summary of the ASTM Methods and Practices for Testing Biodegradation of Polymers

ASTM Code	Purposes	Microorganisms Involved and Key Features	Parameters Monitored	Reference
D5209-92	Aerobic degradation in municipal sewage sludge	Indigenous microorganisms in sewage sludge	CO_2 evolved	[2]
D5210-92	Anaerobic degradation in municipal sewage sludge	Indigenous microorganisms in sewage sludge	CO_2 and CH_4 evolved	[3]
D5247-92	Aerobic biodegradability by specified microorganisms	*Streptomyces badius* ATCC39117 *Streptomyces setonii* ATCC39115 *Streptomyces viridosporus* ATCC 39115 Or other organisms agreed upon	Weight loss, tensile strength, elongation, and molecular weight distribution	[4]
D5271-92	Aerobic biodegradation in activated sludge and wastewater	Municipal sewage treatment plant	Oxygen consumption	[5]
D5338-92	Aerobic biodegradation in conditions composting	2–4 months old compost	Cumulative CO_2 production	[6]
G21-90	Resistance to fungi	*Aspergillus niger* ATCC 9642 *Aureobasidium pullulans* ATCC15233 *Chaetomium globosum* ATCC6205 *Gliocladium virens* ATCC9645 *Penicillum pinophilum* ATCC11797	Visual evaluation	[7]
G22-76	Resistance to bacteria	*Pseudomonas aeruginosa* ATCC 13388	Visual evaluation	[8]

etc., are too large to pass through the cellular membrane. The polymers have to depolymerized to small monomers for absorption in microorganisms. Polymers are broken down aerobically or anaerobically by a variety of enzymes secreted by microorganisms into soil water. The enzymes mainly are intracellular and extracellular depolymerases. First, polymers are broken down into short chains or monomers (depolymerization) using exoenzymes, so that they can permeate through the cell walls. The energy and carbon utilized in this process will end up in products such as carbon dioxide, water, or methane in a process known as mineralization.

The biodegradation process of the biopolymer electrolyte is influenced by humidity, the oxygen level, light, and temperature as well as other factors that enhance the enzymatic degradation of a polymer chain, including:

- The biopolymer electrolyte having amorphous regions with flexible chains inside the matrix of semicrystalline polymer regions and doped salt.
- Chemical bonds (e.g., ester bonds) can be cleaved by the enzyme.
- Functional groups (e.g., —OH, —COOH, —NH_2) easily enter the catalytic center of the enzyme.

6.2 MODE OF BIODEGRADATION [10]

The biological agents such as fungi, bacteria, etc., responsible for the deterioration of polymeric substances with their enzymes are present in the environment. They require suitable conditions of temperature, moisture, and oxygen availability to consume biopolymers as food. The bacteria and fungi are microorganisms of major interest in the biodegradation of natural and synthetic biopolymers.

6.2.1 Fungi

Eumycetes, or true fungi, are nucleated, spore-forming, nonchlorophyllous organisms that reproduce both sexually and asexually. Fungi usually possess filamentous somatic structures and cell walls of chitin and/or cellulose. The fungi suitable for degradation were selected under various testing processes and conditions. The strain of Aspergillus niger is identified by the ATCC Number 9642 or the Quartermaster (QM) Number; the Mycological Services No. 386 was also found to be an acceptable fungi for degradation.

6.2.2 Bacteria

Schizomycetes are bacteria that possess good polymer deterioration capabilities. Soil bacteria are important agents for biopolymer degradation. They can be cocci, rods, spirilla, chain-like, or of a filamentous type. Bacteria can either be aerobic or anaerobic. Some bacteria are chiefly nonchlorophyllous, which depends upon enzymatic degradation of nonliving biopolymer.

6.2.3 Enzymes

Enzymes are biological catalysts that lower the activation energy and thereby increase the reaction rates in the environment. They also require suitable conditions of pH and temperature. Cofactors such as sodium, magnesium, calcium, potassium, or zinc provide optimal activity for the enzyme for degradation. Coenzymes that are organic in nature are usually derived from different B vitamins (thiamine, biotin, etc.). Other coenzymes that are important in metabolic cycles are nicotinamide adenine dinucleotide (NAD^+), nicotinamide adenine dinucleotide phosphate ($NADP^+$), flavin adenine dinucleotide (FAD^+), adenosine triphosphate (ATP), etc. The lock-and-key model explains the absolute specificity of enzymes. The cofactors often induce changes for better orientation of active sites for maximum bonding. The mechanism of degradation varies from a radical mechanism to other chemical routes such as biological oxidation and biological hydrolysis.

6.3 TEST METHODS AND STANDARDS FOR THE BIOPOLYMER ELECTROLYTE

The extent of biodegradation of the biopolymer electrolyte depends on the environment where it has been collected for degradation. The biopolymer electrolyte must be separated from the electrode material and subjected to a complex biological environment such as soil, marine areas, sewage, and compost. The degradation varies from one environment to another because the microorganism will not be able to efficiently degrade selective biopolymers in a single environment. Hence, a standard testing method should be considered for effective degradation of biopolymers after disposal. The American Standard Testing Methods (ASTM) and the Organization for Economic Cooperation and Development (OECD) have proposed several test methods.

The flow chart consisting of the deciding rules is set before a suitable environment is chosen based on the type of biopolymer electrolyte [10]. An example is shown in Fig. 6.1.

6.3.1 Modified Strum Test

This test is the preferred test for biopolymeric materials. The test utilizes aerobic conditions and the sole source of carbon is exposed to low-level inoculums for 28 days without a break in the analysis period. The process of the Sturm method is as follows. To a chemically defined mineral nutrient solution free of organic carbon, the test substance is added at two concentrations (10 and 20 $mg\,L^{-1}$). An inoculum of sewage microorganisms is added ($1–20 \times 10^6\,mL^{-1}$) to the solution. The test system with suitable controls is incubated at ambient temperatures with stirring for 28 days. The CO_2 that evolves is trapped in alkali and measured as carbonate by either titration or with

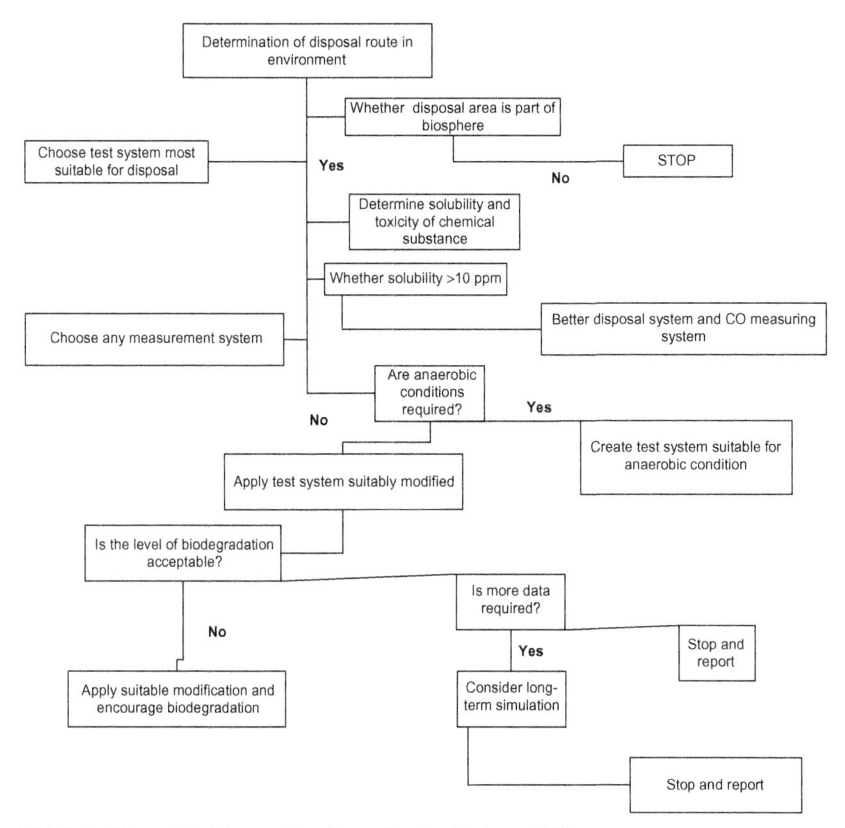

FIG. 6.1 Flow of decision-making for evaluation biodegradability.

the use of a carbon analyzer. After analysis of the data with respect to suitable blank controls, the total amount of CO_2 produced by the polymer over the test period is determined. It is then calculated as that percentage of the total CO_2 that the polymer could have theoretically produced based upon its total carbon content. Because a proportion of carbon will be incorporated into biomass, the total CO_2 and hence the calculated biodegradation levels can never reach 100%.

For a chemical substance to be regarded as readily biodegradable, it should produce >60% of its theoretical total within 28 days. This level should be reached within 10 days of biodegradation reaching 10%.

6.3.2 Closed Bottle Test

The closed bottle test method is as follows. A predetermined amount of the test substance is added to a chemically defined mineral salt solution. The solution is inoculated with sewage microorganisms and then dispersed into closed bottles.

The bottles are incubated in the dark at $20 \pm 1°C$ and periodically assessed for their dissolved oxygen content. The oxygen demand is calculated and compared with the theoretical or chemical oxygen demand of the test substance. Polymers have to be prepared as finely divided powders and their continuous dispersion in the nutrient solution assured. This can only effectively be done using magnetic stirring; this may preclude the use of this test as one test substance requires that at least 25 bottles be used.

6.3.3 Petri Dish Screening Test

This test is used in United States (ASTM), German (DIN), French (AFNOR), Swiss (SN), and international (ISO) standards. The principle of this method involves facing the test material ($2.5 \times 2.5\,cm^2$) on the surface of mineral salts along with agar in a petri dish containing no additional carbon source. The test material and agar surface are sprayed or painted with a standardized mixed inoculum of known fungi or bacteria. The petri dishes are sealed and incubated at a constant temperature between 21 and 28 days. The test material is then examined for the amount of growth on its surface. More growth indicates that the material degrades. The mechanical methods such as weight loss and a stress test by elongating the material can also be carried out after a suitable time of degradation by microorganisms.

6.3.4 Environmental Chamber Method

The environmental chamber employs high humidity (~90%) situations to encourage microbial growth. Strips or prefabricated components of the test materials are hung in the chamber, then sprayed with a standard mixed inoculum of known fungi in the absence of additional nutrients. They are then incubated for 28–56 days at a constant temperature. Visual growth of fungi is observed and reported for degradation.

6.3.5 Soil Burial Test

The biopolymers are buried in soil beds prepared in the laboratory under standard mixtures of compost and minerals. Pretested soils for the degradability of cotton are used for sample burial. The soil beds containing the samples are incubated at a constant temperature for between 28 days and 12 months. The moisture content is normally set at 20%–30%, although it is better calculated as a percentage (40%–50%) of the soil's maximum water-holding capacity. This then accounts for different soil structures and ensures that the soil does not become unduly wet or is too dry for optimal microbial activity. Samples are removed for assessment of changes or a light microscopy and SEM examination to assess surface damage and to look for the presence and nature of microbial growth. Physical factors such as fragmentation and embrittlement can also be assessed in these tests.

6.3.6 Activated Sludge Method

Biopolymers are kept immersed in activated sludge contained in different jars. They are aerated intermittently by air bubbling using a pump to keep the aerobic condition. The films are removed from the sludge solutions on the 5th, 10th, and 15th day by washing with double distilled water. The films are dried in a hot air oven at 75°C and weighed to know the degradation on that 5th, 10th, and 15th day. If weight loss occurs during the process, then the samples are said to be degrading by microorganisms present in the activated sludge treatment. A comparison of several methods used for testing the degradability of various polymers under a range of environment and simulation conditions is shown in Table 6.2.

6.4 SEM ANALYSIS

Scanning electron microscopy (SEM) is a useful imaging approach for the visualization of different polymers because it provides a consistent picture of the polymer morphology as a nonuniform structure characterized by variable thickness and variable polymer density. This technique allows illustrating the surface topography of polymers with high resolution. Due to its high lateral resolution, its great depth of focus, and its facility for X-ray microanalysis (SEM/EDX), SEM is often used in material science—including polymer science—to elucidate the microscopic structure of polymers. In SEM, the surface of nonconductive samples must be coated with a thin layer of gold or platinum. Sometimes, a surface pretreatment (ion sputtering or chemical etching) is carried out to reveal structural details. Moreover, the brittle fracture of samples (in liquid nitrogen in cryo-SEM) can give information about the internal morphology of bulk specimens. SEM micrographs indicate that polymers are characterized by different surface features and the heterogenous local density of chemical components. They also show surface defects such as cracks, etching residues, differential swelling, depressions, and perforations. Currently, a number of different SEM techniques and sample preparation methods have been employed for the study of polymer structure, including ultrahigh resolution field emission SEM (UHR FE-SEM), scanning transmission SEM (STEM), low-vacuum SEM (LVSEM/cryo SEM), and environmental SEM (ESEM). In the LVSEM mode, the delicate polymer samples are observed in the frozen state whereas in ESEM mode, the specimens can contain liquids. An SEM equipped with energy-dispersive X-ray spectrometry profiling (SEM/EDX) is widely used to characterize the variation of the chemical composition of the polymer interface. STEM is used to analyze lamellar arrangements in polymers as well as their dimensions and crystallography. In particular, the recent development of ultrahigh resolution field emission scanning electron microscopy has opened new opportunities in polymer study at the molecular scale [19].

As an example of fungal degradation, peeling and cracking in film texture containing 15% Bionolle in low-density polyethylene (LDPE) were visible

TABLE 6.2 A Comparison of Several Methods Available for Testing Degradability of Different Polymers and Under a Range of Environmental and Simulation Conditions

Methods	Polymer Forms	Inoculums and Degradation Criteria Monitored	Comments	References
Gravimetry	Film or physical intact forms	A wide range of inocula can be used from soil, waters, sewage, or pure species of microorganisms from culture collections	This method is robust and also good for isolation of degradative microorganisms from an environment of interest. Reproducibility is high. Disintegration of a polymer cannot be differentiated from biodegradation	[11]
Respirometry	Film, powder, liquid, and virtually all forms and shapes	Either oxygen consumed or CO_2 produced under aerobic conditions. Under methanogenic conditions, produced methane can be monitored	This method is most adaptable to a wide range of materials. It may require specialized instruments. When fermentation is the major mechanism of degradation, this method underestimates the results	[12]
Surface hydrolysis	Films or others	Generally aerobic conditions, pure enzymes are used. Hydrogen ions (pH) released are monitored as incubation progresses	Prior information about the degradation of the polymer by microorganisms or particular enzymes is needed for the target-specific test	[13]
Electrochemical impedance spectroscopy (EIS)	Films or coatings resistant to water	The test polymers should adhere on surface of conductive materials and electrochemical conductance is recorded	Polymer must be initially water impermeable for signal transduction. Degradation can proceed quickly and as soon as degradation is registered, no further degradation processes can be distinguished	[14–18]

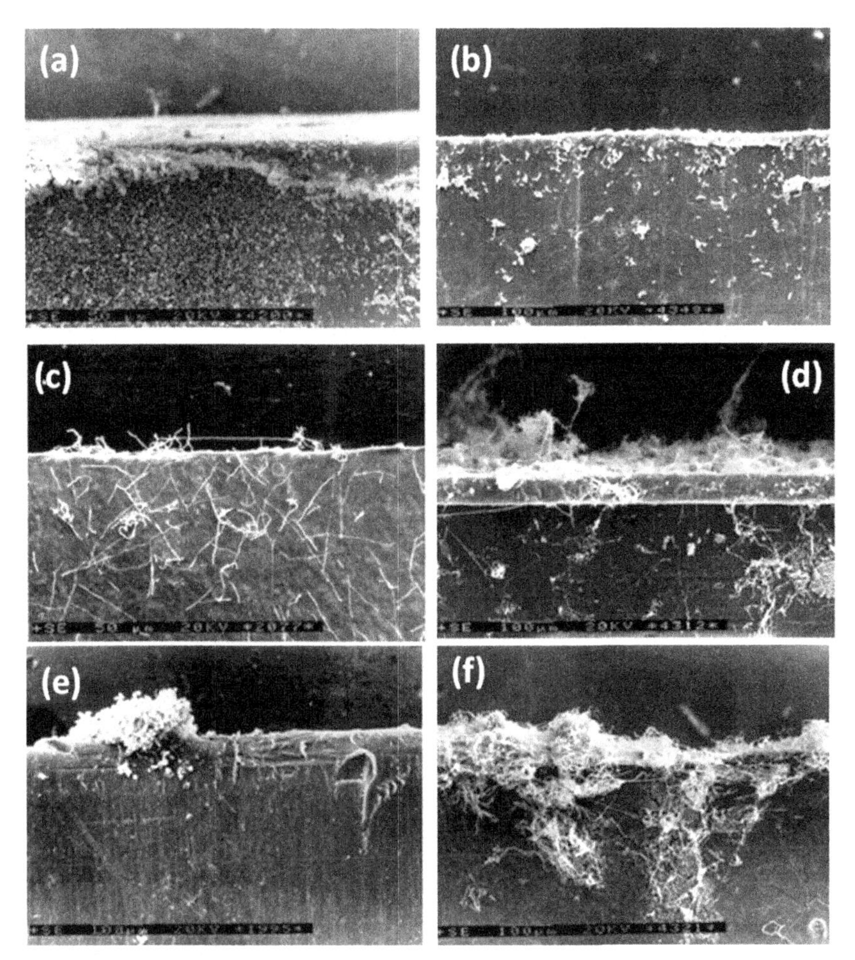

FIG. 6.2 SEM micrographs of neat LDPE/Bionolle 85/15 composition after biodegradation with (A) *Aspergillus niger,* (B) *Aureobasidium pullulans,* (C) *Paecilomyces varioti,* (D) *Penicillium funiculosum,* (E) *Trichoderma viride, and* (F) mixed fungal population [19].

(Fig. 6.2). The entire surface of the film was densely covered with spores belonging to *Aspergillus niger* (Fig. 6.2A) and *Aureobasidium pullulans* (Fig. 6.2B). Scarce hyphae of *Paecilomyces varioti* (Fig. 6.2C), *Penicillium funiculosum* (Fig. 6.2D) and a mixed population of fungi (Fig. 6.2F) colonized both the edges and surface of the film. Agglomerations of *Trichoderma viride* conidiophores (Fig. 6.2E) primarily inhabited the edges of the sample.

After an 84-day incubation with filamentous fungi, the most intense changes were found on the surface of the 40/60 LDPE/Bionolle composition. Hyphae and conidiophores of *Aspergillus niger* (Fig. 6.3A), *Aspergillus terreus*

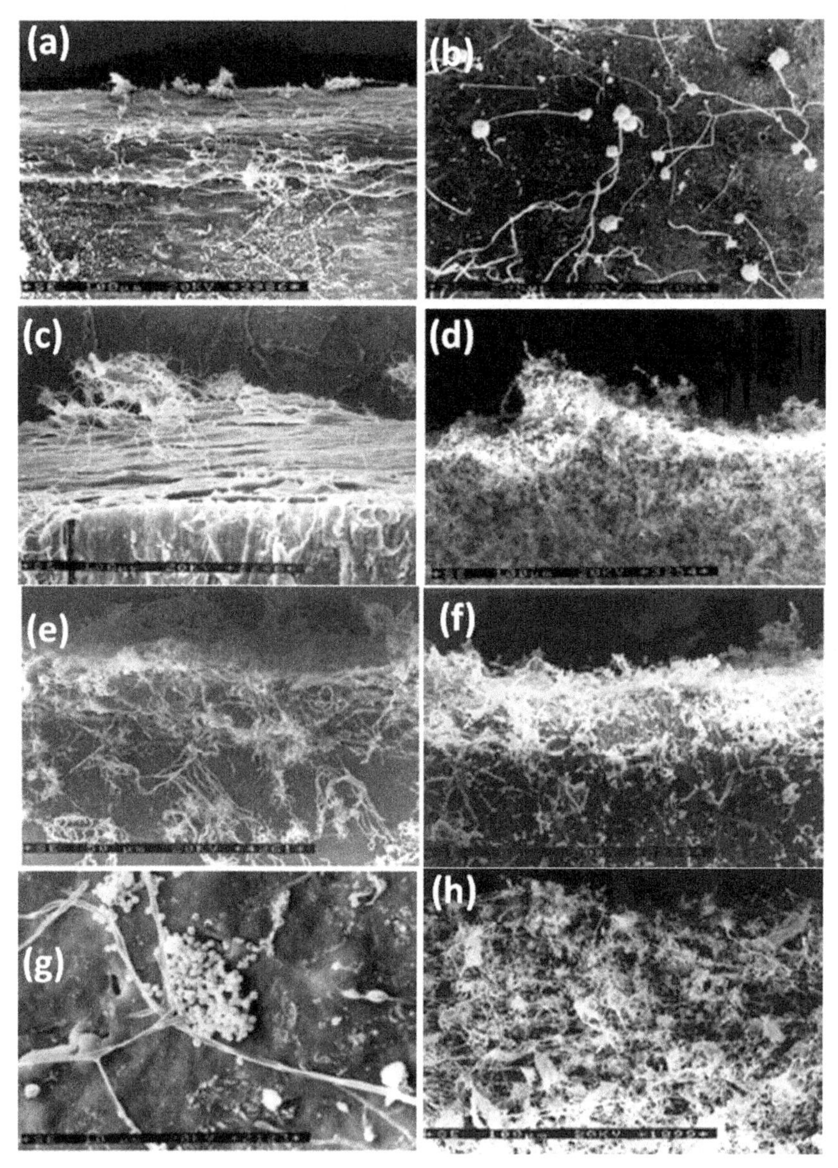

FIG. 6.3 SEM micrographs of neat 85/15 composition after biodegradation with (A) *Aspergillus niger*, (B) *Aureobasidium pullulans*, (C) *Paecilomyces varioti*, (D) *Penicillium funiculosum*, (E) *Trichoderma viride*, and (F) mixed fungal population [19].

(Fig. 6.3B), *Aureobasidium pullulans* (Fig. 6.3C), and *Trichoderma viride* (Fig. 6.3G) grew out directly from the polymer sample. However, changes induced by the action of these microorganisms were deep and distinct. A dense network of fractures was particularly visible after incubation with *Aspergillus terreus*.

Aspergillus niger and *Aureobasidium pullulans* caused massive exfoliation of the plastic edges. Although *Penicillium ochrochloron* (Fig. 6.3E) and *Scopulariopsis brevicaulis* (Fig. 6.3F) created mycelium, a considerable part of the film surface was found to be unaffected. A loss of film integrity resulted in the fragile surface being entirely covered by dense mycelia of *Penicillium funiculosum* (Fig. 6.3D) and a mixed population of fungi (Fig. 6.3H).

In the environment, plastics decompose under the influence of different abiotic and biotic factors. The abiotic factors such as radiation, temperature, humidity, chemical pollution, and wind can act synergistically or antagonistically, causing various types of structural and chemical changes in the polymer. Microorganisms, especially bacteria or fungi, play a crucial role in the biological degradation of polymers.

REFERENCES

[1] Dey U, Mondal NK, Das K, Dutta S. An approach to polymer degradation through microbes. IOSR J Pharm 2012;2(3):385–8.

[2] ASTM (American Society for Testing and Materials). 1993 Annual book of ASTM standards, vol. 08.03, vols. 5209-92, Philadelphia, Pennsylvania; 1993. p. 377–80.

[3] ASTM (American Society for Testing and Materials). 1993 Annual book of ASTM standards, vol. 08.03, D5210-92, Philadelphia, Pennsylvania; 1993. p. 381–84.

[4] ASTM (American Society for Testing and Materials). 1993 Annual book of ASTM standards, vol. 08.03, D5247-92, Philadelphia, Pennsylvania; 1993. p. 401–04.

[5] ASTM (American Society for Testing and Materials). 1993 Annual book of ASTM standards, vol. 08.03, D5271-92, Philadelphia, Pennsylvania; 1993. p. 411–16.

[6] ASTM (American Society for Testing and Materials). 1993 Annual book of ASTM standards, vol. 08.03, D5338-92, Philadelphia, Pennsylvania; 1993. p. 444–49.

[7] ASTM (American Society for Testing and Materials). 1993 Annual book of ASTM standards, vol. 08.03, G21-90, Philadelphia, Pennsylvania; 1993. p. 527–29.

[8] ASTM (American Society for Testing and Materials). 1993 Annual book of ASTM standards, vol. 08.03, G22-76, Philadelphia, Pennsylvania; 1993. p. 531–33.

[9] Gu J-G, Gu J-D. Methods currently used in testing microbiological degradation and deterioration of a wide range of polymeric materials with various degree of degradability: a review. J Polym Environ 2005;13(1):65–74.

[10] Chandra R, Rustgi R. Pergamon biodegradable polymers. Prog Polym Sci 1998;23 (97):1273–335.

[11] Luckachan GE, Pillai CKS. Biodegradable polymers- a review on recent trends and emerging perspectives. J Polym Environ 2011;19(3):637–76.

[12] Gu JD, Gada M, Kharas G, Eberiel D, McCarty SP, Gross RA. Polym Mater Sci Eng 1992;67:351–2.

[13] Gu JD, McCarty SP, Smith GP, Eberiel D, Gross RA. Polym Mater Sci Eng 1992;67: 230–1.

[14] Kemnitzer JE, McCarty SP, Gross RA. Macromolecules 1993;23:6143–50.

[15] Gu JD, Ford T, Thorp K, Mitchel R. Int Biodeterior Biodegrad 1996;39:197–204.

[16] Gu JD, Ford T, Thorp K, Mitchel R. J Appl Polym Sci 1996;62:1029–34.

[17] Breulmann M, Künkel A, Philipp S, Reimer V, Siegenthaler KO, Skupin G, et al. Polymers, biodegradable. In: Ullmann's encyclopedia of industrial chemistry. Weinheim: Wiley-VCH; 2009.

[18] Rudnik E. Biodegradability testing of compostable polymer materials. In: Handbook of biopolymers and biodegradable plastics. USA: Elsevier; 2013. p. 213–63.

[19] Bożena N, Jolanta P, Jagna K. Biodegradation of pre-aged modified polyethylene films. In: Kazmiruk V, editor. Scanning electron microscopy. InTech; 2012. https://doi.org/10.5772/35128.

Index

Note: Page numbers followed by *f* indicate figures and *t* indicate tables.

G

H

I

L

U

V

W

X

Printed in the United States
By Bookmasters